Toxic Contamination in Large Lakes

Volume II Impact of Toxic Contaminants on Fisheries Management

Edited by Norbert W. Schmidtke

Library of Congress Cataloging-in-Publication Data

Impact of toxic contaminants on fisheries management.

(Toxic contamination in large lakes ; v. 2)
Proceedings of a technical session of the World Conference on Large Lakes, held May 18–21, 1986, Mackinac Island, Mich.
Includes bibliographies and index.
1. Fishes—Effect of water pollution on—Congresses.
2. Fishery management—Congresses. I. Schmidtke, N. W. II. World Conference on Large Lakes (1986 : Mackinac Island, Mich.) III. Series.
QH545.W3T67 vol. 2 363.7′394 s 87-37940
[SH174] [333.95′6153]
ISBN 0-87371-090-8

LEWIS PUBLISHERS, INC.
121 South Main Street, Chelsea, Michigan 48118

PRINTED IN THE UNITED STATES OF AMERICA

PREFACE

The publication of **Toxic Contamination in Large Lakes**, a four-volume set of resource books, is the proceedings from the 1986 World Conference on Large Lakes. Governor James J. Blanchard and the State of Michigan hosted this international conference on Mackinac Island, May 18-21, 1986.

The proceedings represent research findings delivered at one of the first world conferences where government officials, policy makers, citizens, scientists, and business leaders joined together to explore solutions to toxic contamination in the large lakes.

Unique to the world conference was the emphasis on an integrated approach to examining toxic pollution in large lakes. The Conference covered not only scientific research, but also social and political concerns.

Over 500 world experts took this opportunity to share the latest research, management techniques, and political strategies for remedying large lakes from toxics.

The objectives at the Conference were:

1. Define the state of scientific knowledge with respect to toxic substances;

2. Identify successful, preventive, regulatory, and management alternatives;

3. Explore and encourage public education and involvement in lake management.

The primary concern of the conference participants was the fact that migrating toxic contaminants ignore political boundaries and can only be effectively controlled through international cooperation. Many of the discussions, as reflected in these proceedings, focussed on three key points concerning transboundary relationships in the management of the world's large lakes.

First, to resolve the problem of toxic contaminants requires a commitment from well-informed policy makers, scientists, citizens, and business leaders from all involved jurisdictions.

Second, jurisdictions must look beyond the political boundaries and study entire watersheds to establish prevention and control measures. New studies showing that airborne pollutants travel far greater distances than previously considered indicate the need for **ecosystem management** on a global basis.

Third, natural media - air, soil, and water - need to be managed by a cross-media approach.

These three points are well explored in the proceedings and deliberated upon in the conference synopsis that is featured in this publication.

Overall, the Conference represented a turning point for the future cooperation in protecting the world's large lakes. It brought the threat of toxic contamination to large lakes to the forefront of international environmental issues.

The documentation of this most important informal exchange will serve as a valuable reference to researchers, governments, business, and citizens around the world in their shared efforts to protect the earth's greatest freshwater resources.

William D. Marks
Chairman
World Conference on Large Lakes
Mackinac '86

ACKNOWLEDGMENTS

Special recognition is given to General Motors Corporation for underwriting the editing and copy preparation of the proceedings.

Other organizations that generously provided resources for the 1986 World Conference on Large Lakes and the publishing of the proceedings are: Allied-Signal, Inc.; Dow Chemical USA; Electronic Data Systems, Inc.; Hiram-Walker-Gooderham & Worts, Ltd.; The International Joint Commission; The Joyce Foundation; W.K. Kellogg Foundation; Mead Corporation; Michigan Department of Natural Resources; Michigan Department of Commerce; Michigan Sea Grant College Program; Charles Stewart Mott Foundation; Shell Western E & P; The Stroh Brewery Company; The Upjohn Company; and the United States Environmental Protection Agency-Great Lakes National Program Office.

The Shiga Prefecture of Japan is commended for originating the biennial world conference on large lakes in 1984. It was through the Shiga Prefecture's invitation to the State of Michigan to host the second World Conference on Large Lakes, that Mackinac '86 came to be.

Norbert W. Schmidtke is a consulting engineer and internationally recognized authority on wastewater treatment technology with 25 years experience in research, development, design, application and technology transfer. Prior to returning to consulting practice Dr. Schmidtke was director of Canada's Wastewater Technology Centre.

Dr. Schmidtke holds B.Sc. and M.Sc. degrees from the University of Alberta and a Ph.D. from the University of Waterloo. All degrees are in civil engineering. He is a registered professional and consulting engineer in the Province of Ontario, diplomate of the American Academy of Environmental Engineers and member of WPCF, PCAO, CAWPRC, IAWPRC, AEEP, CSCE and AIDIS. He is also associate editor (Environmental Division) of the Canadian Journal of Civil Engineering.

His areas of expertise range from industrial and municipal wastewater treatment process selection and analysis, to aeration technology, eutrophication, sludge and hazardous waste management. His special interest concerns process scale-up.

Dr. Schmidtke consults to industry, consultants, WHO and government organizations nationally and internationally. He has organized and is principal lecturer in many continuing education courses and custom-designed in-house training programs to industry, academia and governments worldwide. He was the 1981/82 national lecturer of the Canadian Society for Civil Engineering and recently represented CSCE on a lecture tour in the People's Republic of China. He is academically active through adjunct appointments in civil and chemical engineering.

Dr. Schmidtke has served as chairperson for many expert technical, advisory committees and Task Forces and recently chaired the Natural Sciences and Engineering Research Council's (NSERC) Grant Selection Committee for civil engineering.

Dr. Schmidtke has authored more than 130 technical papers and reports, and recently co-edited: 'Scale-Up of Water and Wastewater Treatment Processes'.

Four Volume Proceedings Author Index

Contents

VOLUME II

IMPACT OF TOXIC CONTAMINANTS ON FISHERIES MANAGEMENT

Contents

WORLD CONFERENCE ON LARGE LAKES - MACKINAC '86

MAY 18-21, 1986

MACKINAC ISLAND, MICHIGAN, USA

Hosted by: JAMES J. BLANCHARD
GOVERNOR, STATE OF MICHIGAN

FOREWORD

The World Conference on Large Lakes brought together scientists, policy makers, business and citizen leaders to discuss issues associated with toxic contamination. Significant issues confronting the quality of the world's large lakes were chosen as conference themes and explored during four concurrent sessions:

-Chronic Effects of Toxic Contaminants in Large Lakes

-Impact of Toxic Contaminants on Fisheries Management

-Sources, Fate and Controls of Toxic Contaminants

-Prevention of Toxic Contamination of Large Lakes

The complete record of the technical presentations is contained in a 4 volume Proceedings set. In addition, the co-conveners for each technical session prepared a summary including findings and conclusions of their session. These summaries were designed to provide conference participants and other interested parties with a synopsis of the presentations and discussions at Mackinac '86. The co-conveners then used their summary to develop specific recommendations.

Each of the technical session's presentations, complete with session summary and recommendations, is presented in its separate Conference Proceedings volume. Volume 4 contains all session summaries and recommendations.

More than 500 people actively participated in a dynamic and productive exchange of ideas where almost 100 presentations were made. Conference planners and staff are grateful to all those who contributed their expertise and enthusiasm in this unique forum. We hope that the conference has been valuable to all who attended, and that relationships established here at Mackinac Island will support a growing effort to effectively prevent toxic contamination of large lake ecosystems worldwide.

A brief introduction to the topic of toxic contaminants in large lakes and conference summary follows.

INTRODUCTION

Large lakes contain a significant portion of the earth's water. For example, there are 253 large lakes greater than 500 km^2 which have a total surface area of 1,456,149 km^2 and an estimated volume of 202,000 km^3. Collectively, these 253 large lakes account for 93 and 88%, respectively, of the total surface area and volume of water held in all lakes of the world. These lakes, if only by virtue of the water they hold, constitute the world's most valuable resource. Large lakes fulfill key roles in the economy and overall well-being of humankind. They are used for drinking water, recreation, fishing, transportation, cooling water, and waste assimilation. Some large lakes, such as Lake Maracaibo in Venezuela are important sources of minerals and petroleum. Many large lakes, like the Great Lakes in North America, are particularly valuable as a source of drinking water and as a source of water for manufacturing. Large lakes, such as those in East Africa, also may be a life sustaining source of irrigation water.

Large lakes are also important for their ecology. The variety of large lake habitats result in a great species diversity. For example, Lake Baikal in the USSR, the world's deepest lake (maximum depth: 1,741 m), supports approximately 1,700 biological species, of which 1,200 are found nowhere else. Such species diversity is not only unparalleled, but of fundamental importance in maintaining the genetic pool and uniqueness of that aquatic ecosystem.

Many large lakes have suffered from overfishing, destruction of biological habitat (e.g. wetlands, spawning and nursery grounds for fishes), cultural eutrophication, or toxic substances pollution. In 1984, the first World Conference on Large Lakes Environment was held in Shiga, Japan. The theme of the 1984 conference was eutrophication. Michigan Governor James Blanchard accepted an invitation from Shiga (Michigan's sister state) to host the second conference focusing on toxic substances in large lakes.

Many toxic substances are highly persistent and can bioaccumulate in fishes and wildlife to levels which threaten aquatic ecosystems and human health. Large lakes, because of their long residence and flushing times (e.g. Lake Superior's flushing time is 181 years), are particularly susceptible to persistent toxic substances. The input of persistent toxic substances such as polychlorinated biphenyls (PCBs) and DDT has rapidly resulted in long-term, whole-lake problems in many large lakes.

Toxic substances can enter large lakes via manufacturing processes (e.g. industries), use (e.g. agriculture), and disposal practices (e.g. landfills). It is now clear that the atmosphere is a major pathway by which synthetic organic contaminants and trace metals enter large lakes. Long-range transport of toxic substances via the atmosphere is contributing to contamination of most large lakes throughout the world.

The 1986 World Conference on Large Lakes has brought together world experts on toxic substances in large lakes and key policy makers from jurisdictions which include large lakes. This conference has provided an opportunity for representatives of industrialized nations to share information on sources of toxic pollution, effects, and viability of control programs used to date. Representatives of less developed nations have had an opportunity to learn from the experiences of more developed nations. Policy makers have benefited from the opportunity to discuss transboundary problems, such as long-range transport of toxics through the atmosphere and multimedia problems, such as movement of toxic substances from landfills to lakes via groundwater seepage. The conference has also provided an avenue for concerned citizen leaders to share effective mechanisms for working with governments and business to control toxic substance pollution of large lakes.

CONFERENCE SUMMARY

The 1986 World Conference on Large Lakes has confirmed the crucial need for the most basic of commodities - **information.** Governments and public institutions have a responsibility to communicate to citizens and the media clear, concise, and accurate data concerning the environmental health and well-being of the ecosystem. For example, the public deserves more complete information and criteria regarding fisheries closures or advisories. Human exposure and chronic health effects must be documented relative to long-term exposure to multiple persistent toxic chemicals.

With regard to human health, the economics of the fisheries, and the overall health of the ecosystem, the solution is also simple - **stop the input of contaminants.** The world is well aware that attention should continue to be placed on the control of known point sources. As an example, the report of the US-Canadian International Joint Commission on the role of non-point pollution in North America has served as a strong basis for the development of new strategies to cope with non-point source pollution. However, the focus of most studies remains limited to conventional pollutants. The next step for the world community is to focus on the control of toxic substances for both point and non-point sources and to monitor for emerging pollutants.

Historically, scientific inquiry and public policy development have been regarded as mutually exclusive elements in the management of large lake systems. A research brokerage function, designed to link science and public policy is largely absent in large lake management systems.

It is not enough that the scientific community continues to recognize the interrelationships of our environment through an **"Ecosystem Approach".** The regulatory community, through political processes, must respond by implementing policy guided by a perspective of our interrelated environment which extends beyond national boundaries or environmental compartments and must arrange their institutions accordingly. The world community must adopt a philosophy of prevention of toxic substance contamination rather than merely reacting to environmental crises.

CHAPTER 1

IMPACT OF TOXIC CONTAMINANTS ON FISHERIES MANAGEMENT

Technical Session Co-Conveners:

Niles Kevern[1] and H. Francis Henderson[2]

[1] Associate Director, Michigan Sea Grants College Program, Michigan State University, East Lansing, MI
[2] Chief, FIRI, Food and Agriculture Organization of the United Nations, Rome, Italy

As aquatic ecologists, limnologists, and fisheries scientists and managers, we are often frustrated in our efforts to convince the public and decision makers of the importance of our large lakes or of our aquatic resources in general. Narrowing the subject to contaminants or toxic substances at this conference did not alleviate this frustration. However, several points seemed to be common among the presentations and addressing these points as conclusions and translating them into recommendations may help relieve the frustrations and lead to more rapid progress in our efforts to maintain the integrity and value of our large lakes.

Most of us feel that we must use an ecosystem approach in understanding and managing our large lakes. At the same time, we realize that this is the most difficult approach. It is the most difficult for the public to understand and it is the most difficult to institutionalize, as it requires a high degree of integration and cooperation. So, what must we do to achieve this ecosystem approach? We must of course, continue our efforts to work in that direction and, although time is precious, we must be patient and persistent.

How do we gain public support for decisions that favor our large lakes? How do we impress the public with the value of these our lakes so that they, in turn, will support, even demand, our law makers to make the right decision? The key term is value. Our publics appreciate more readily those aspects that touch upon them directly. In this instance, they appreciate their water supply and they can appreciate their food, the fish they eat. They less readily understand constants governing solution or vaporization of chemicals, partitioning of organic compounds into fats and oils, or the phytoplankton and zooplankton interactions that are part of the aquatic foodweb. It is easier to understand that 24 million people rely on the Laurentian Great Lakes for their drinking water and that 13 million people rely on Lake Biwa in Japan for their drinking water. And that you get thirsty very quickly when your water supply is gone or that it would be extremely costly to remove all contaminants from seriously contaminated water.

It is easier to appreciate the lakes when we realize that the annual economic value of the salmonid fishery in the Great Lakes is 2-3 billion dollars and that the dollar value of freshwater pearl production in Lake Biwa is significant. We must ask our resource economist to continue to research and determine the worth of our large lakes. We must ask our social scientists to research and determine the non-market values of our lakes as perceived by the public. Hopefully we can realize a value high enough to convince decision makers to support additional research into subjects that require longer periods of time.

Several times in our session we noted that the public is often confused by our information on toxic substances in fish. We blamed the public media, but we also realized that we shared the blame. We have not pooled and coordinated our own information within or among our agencies. No wonder the public is confused. We tell them that eating fish is good to prevent heart disease and then issue health advisories about not eating fish because of contaminants. We need to do better in coordinating information among state, provincial and federal governments. In North America, the Great Lakes Fishery Commission (GLFC) is increasingly serving as a coordinating body among the state and provincial fishery management agencies. The GLFC and similar organizations elsewhere in the world need to place an even greater emphasis on this coordinating role. We must be more conscious of the need to present complete information to the media and when possible, to insist on more complete transferal to the public. The media must be aware of the very great role that they play in public education.

Part of our failure to get complete information to the public is because we are still missing information. We need to monitor our programs in contaminant control better. Much of our data to date are not standardized, leading to great variability and thus non-comparability. For monitoring purposes, we should standardize the species of fish (perhaps the lake trout), the size of the fish, the time of the year, the method of preparation, the analytical method, etc. We wouldn't remove all variability, but we could improve the comparability.

We also lack information on the real effects of contaminants on human health and on aquatic ecosystem health, thus we need long-term research on both aspects. Tolerance levels for contaminants in humans and in aquatic systems and action levels for fish must be based on continually improving data and on long-term, sub-lethal effects. Risk assessments are unreal unless such data are available. Realistic management of fish stocks is impossible without such data. For aquatic systems, we need microcosm or mesocosm approaches where the laboratory study becomes as natural as possible. Long-term population predictions for humans or fish are not realistic without data on carcinogen or mutagen effects.

In our session, we often observed a conflict between human health and economic values when establishing action levels of a contaminant for fish. This conflict, for lack of better information, leads to a judgmental action level. Ideally, with regard to human health, economics of the fishery and health of the ecosystem, the solution is really simple - stop the input of the contaminant. We watched the recovery of Lake Erie and heard about the beginning of the recovery of Lake Orta in Italy when nutrients or contaminant inputs to the lakes are stopped. We know that recovery can be good in most cases when we stop the input. We have done reasonably well at treating point sources to remove contaminants, but yet some point sources seem to elude us. We must bring increasing pressure to bear on known point sources and continue to identify unknown points. Increased effort is needed to reduce non-point input of known contaminants.

Again and again, speakers mentioned the vital role of public involvement. When the publics are convinced of the value of the resource, they will support the protection of that resource. We must convince them of the value of our large lakes.

RECOMMENDATIONS

1. More research is needed to determine the value, both economic and social, of our large lakes.

2. A responsible organization is needed to coordinate information for the public on toxics in fish.

3. More research is needed to establish realistic, long-term tolerance levels for fish health, aquatic ecosystem health, using micro-and mesocosms, and for the health aspects related to human consumption of fish.

4. Standardization of monitoring contaminants in fish is needed for comparing long-term trends and an early warning system.

5. Continuing efforts are needed to reduce or stop inputs of contaminants to large lakes from both point and non-point sources.

6. Develop a procedure to bring the public, possibly through advisory groups, into working groups with scientists and resource managers to address contaminant problems.

CHAPTER 2

FISHERIES MANAGEMENT, WATER QUALITY AND ECONOMIC IMPACTS: A CASE STUDY OF LAKE KINNERET

M. Gophen

Yigal Allon Kinneret Limnological Laboratory, P.O.B. 345, Tiberias 14102, Israel

ABSTRACT

Lake Kinneret, Israel, currently supplies 35% (400-430 x 10^6 m^3/year) of the national water consumption. Of this, the proportion supplied for domestic use is increasing rapidly because of larger demands. The lake is also utilized for fisheries (1800 tons/year, i.e. 106 kg/ha), recreation and tourism. Although the top national priority for lake utilization is water supply, fisheries, especially stocking management policy, affect water quality. Two-thirds of the fishery income is due to stocked species, of which 40%, 35%, 15% and 10% came from catches of grey mullet, St. Peter's fish, silver carp and blue tilapia respectively. Recently zooplankton biomass decreased, the densities of small cladocerans relative to large ones increased, and a reduction in Bosmina body size was observed. Daphnia lumholtzi disappeared from lake Kinneret. Consequently, there was an intensification of fish predation on zooplankton, the major food source in summer. As a result, the biomass of nanoplankton increased, causing water quality deterioration. Introduced fingerlings contributed 150-320 and 85-150 kg (catch)/10^3 fingerlings/year during 1954-1962 and 1962-1984, respectively. Stock-catch correlation values were high (r^2 = 0.78) when introduction varied between 0.5 - 4.0 x 10^6 fingerlings/year of all stocked fish. Fishery management recommendations aimed at prevention of water quality deterioration, as well as sufficient income, are: elimination of blue tilapia and silver carp stocking; annual introduction of 5×10^6 and 1×10^6 fingerlings of St. Peter's fish and grey mullet, respectively; intensification of bleak exploitation; and prevention of criminal utilization of pesticides for fishery.

INTRODUCTION AND BACKGROUND

Lake Kinneret, the only natural freshwater lake in Israel, currently supplies 35% of the national water consumption. In Israel the major water sources are located in the northern part of the country, whereas most of the populated and cultivated areas are in the central and drier southern regions, respectively. The principal component of the water supply system, therefore, is the National Water Carrier (NWC), which pumps water from Lake Kinneret to the central and southern areas. The NWC system was put into operation in 1964.

Prior to the operation of the NWC system, priorities of lake utilization were different. At present, water supply is the top national priority, whereas before 1964 it was fisheries. During the period 1964-1984, 6×10^9 m^3 of water were pumped by the NWC, which is about 1.5 times the total lake volume. During the winter season, pumped lake water is utilized mostly for drinking and stored in underground aquifers along the coastal plain. During spring-summer-fall seasons, NWC water is mostly used for drip irrigation, as well as domestic use. The quality standards of water supplied for drinking, underground infiltration and drip irrigation are very strict: low concentrations of suspended matter and organics, low salinity, and in drinking water the absence of pathogenic bacteria, are ultimately needed.

The water quality of Lake Kinneret is of national concern because it is proposed in the future to utilize this water body as the major drinking water source for the country.

To prevent water quality deterioration in Lake Kinneret, two principal types of activities were proposed:

1. reduction of external inorganic and organic nutrients, including salt influx from salt springs, and

2. improving policies of fisheries and stocking.

The first proposal has already been largely implemented with the help of heavy financial investments by the national water authority and local farming organizations.

At present, the standard quality required for drinking water in Israel makes it possible to utilize Kinneret water with only minor chlorination. There is no need for further treatment if the municipal water supply systems are operated properly. Utilization of Kinneret water for drinking without complicated and expensive treatment, is a result of the management policy for the watershed area and the lake itself. About 60% of the nutrients originating from domestic sewage and fish pond wastes were removed to reservoirs for reuse. One of the reasons for the advanced implementation of the external nutrient reduction was the good cooperation between the Kinneret limnologists, the water authority (Water Commission and Kinneret Authority) and the Farmer's Organization. This permitted not only efficient utilization of the limited resources but also proper management of the new system and solving land ownership difficulties for the reservoir construction. Implementation of a fisheries policy in Lake Kinneret is also required, since food web alterations originating from increasing fish planktivory may diminish water quality improvement caused by external nutrient reduction.

Shapiro and co-workers were among the first scientists to recognize the potential of biomanipulation as a tool for food web alterations and water quality impact (Shapiro 1980, Shapiro and Wright 1984). Stephen et al. (1985) analyzed food web alterations by modifying consumer populations and also suggested the promising significance of this management tool. Fishery management based on logistic models and their ecological impact, nutrient recycling by zooplankton, and the strong impact of fish selection predation on zooplankton, have been described (Bartell and Kitchell 1978, Larkin 1978, Kitchell 1980, Lehman 1980, Drenner et al. 1982a, Shapiro et al. 1982, Stephen et al. 1984). Fish introduction, especially exotic fish, may provide a rich food source but not always without ecological cost (Zaret and Paine 1973). Barel et al. (1985) published a dramatic commentary concerning the destruction of fisheries in African lakes, such as the introduction of the Nile perch into Lake Victoria, which had disastrous consequences.

The direct and indirect impact of fisheries and stocking policies on the ecological structure and water quality of Lake Kinneret has been widely investigated (Gophen and Landau 1977, Serruya et al. 1979, 1980, Gophen 1980, 1984, 1985a,b,c, 1986, Gophen and Scharf 1981, Drenner et al. 1982b, Gophen et al. 1982 a, b, c, Gophen and Pollingher 1985). Fisheries management and stocking policy in Lake Kinneret have recently been under intensive discussion. The administrative structure of the organizations responsible for the utilization of Lake Kinneret fisheries is complicated and not without contradictions. The long-term debate between fishery biologists and limnologists (Rigler 1982), concerning fisheries management is also applicable to Lake Kinneret.

It was difficult, however, to include fisheries regulations in the overall program of Kinneret water quality improvement. Fisheries management is based on laws which were formulated and officially implemented when the lake was not utilized as the major source of drinking water for Israel. From 1964, when the NWC became operational, laws and policy for fisheries management remained in force, although limnological conclusions aimed at water quality improvement suggested that modifications were necessary.

In this paper new information regarding aspects of fisheries impacts on water quality and the economy are presented.

FISHERIES AND STOCKING ACTIVITIES

Tables 1 and 2 summarize the fishery and stocking activity data for the periods 1967-1984 and 1960-1984 respectively (all lake fishery resources [fishing areas] were exploited by Israeli fisherman and monitored by the Fishery Department, Ministry of Agriculture, after 1967). During this period the total annual catches were relatively stable and averaged 1800 ± 205 tons (106 kg/ha) (Sarid 1950-1979, Golani 1980-1984), one of the highest values known for deep lakes (Fernando and Holcik 1982).

During this period, the number of licensed fishermen and boats did not change very much. On the other hand, there were significant changes in fishing effort: net mesh size decreased, length and height of nets increased, and many sonar systems to locate fish schools were introduced.

TABLE 1 FISHERY ACTIVITIES IN LAKE KINNERET FOR THE PERIOD 1967 - 1980. (Sarid 1950-1979, Golani 1980-1984).

NAME	AVERAGE tons	CATCH %	COMMERCIAL VALUE	REMARKS
Bleaks:				
<u>Mirogrex terraesanctae</u> <u>Acanthobrama lissneri</u>	990	55.0	low	endemic
Silver carp	130	7.2	medium	exotic
Barbels:				
<u>Barbus longiceps</u> <u>Tor canis</u> <u>Capoeta damascina</u>	131	7.3	medium	
Cichlids:				
<u>Tilapia zillii</u>	3	0.2	low	
<u>Sarotherodon galilaeus</u>	220	12.2	high	introduced
<u>Oreochromis aureus</u>	80	4.4	high	introduced
<u>Tristramella simonis</u> <u>Tristramella intermedia</u> <u>Tristramella sacra</u>	42	2.3	medium	very rare
Grey mullet:				
<u>Mugil cephalus</u> <u>Liza ramada</u>	200	11.1	high	exotic
Catfish	4	0.3	low	
TOTAL	1800	100.		

TABLE 2 STOCKING ACTIVITIES IN LAKE KINNERET FOR THE PERIOD 1960-1984. Numbers are total stocked fingerlings in 10^6 (Sarid 1950 - 1979, Golani 1980 - 1984).

SPECIES	NUMBER OF FINGERLINGS X(10^6)
<u>Sarotherodon galilaeus</u>	38 (34%)
<u>Oreochromis aureus</u>	32 (29%)
Silver carp	13 (12%)
Grey mullet	29 (25%)
	TOTAL: 112

There are fifteen commercial fish species in Lake Kinneret, of which two are endemic (bleaks), three are exotic (grey mullets and silver carp), and *Oreochromis aureus* was very rare prior to its introduction. Catfish (*Clarias gariepinus*) and *Tilapia zillii* are fished in very low quantities due to their low commercial value. The most abundant fish in Lake Kinneret (Gophen and Landau 1977) are the endemic bleaks (*Mirogrex* sp. and *Acanthobrama* sp.). Annual catches of bleaks are 55-60% of the total but their commercial value is low. Stocked fish are the native *Sarotherodon galilaeus* (Galilee St. Peter's fish), *O. aureus* (Jordan St. Peter's fish, blue tilapia), grey mullet and silver carp. The proportion of stocked fish in the annual catches varied between 23-38% (excluding the exceptional year of 1983), but since their market demands are highest, their commercial value is approximately 66% of the total fishery income from Kinneret fish (Table 3).

In order to find the contribution of stocking activities to annual catches the following parameters were calculated for all introduced species (*S. galilaeus*, *O. aureus*, grey mullet and silver carp):

$$I_i = \frac{I(i-1) + I(i-2)}{2}$$

$$T_i = \frac{\text{Annual catches}}{I_i}$$

where:

I_i = average number of stocked fingerlings
$I_{(i-1)}$ = numbers of stocked fingerlings 1 year before i
$I_{(i-2)}$ = numbers of stocked fingerlings 2 years before i
T_i = annual catches (kg) of stocked species per 10^3 introduced fingerlings i.e. index of stocking success

To correlate stocking to catch, an average of 1 and 2 years prior to year i was taken (Ben-Tuvia et al. 1983).

T_i and I_i values were smoothed by averaging 3 consecutive years. Results of these running averages are presented in Figures 1 and 2.

Two different ranges of catch per 10^3 introduced fingerlings are clearly shown in Figure 1: 150-320 and 85-150 kg (catch)/10^3 fingerlings/year, during 1954-1962 and 1963-1984, respectively. Correlation analysis

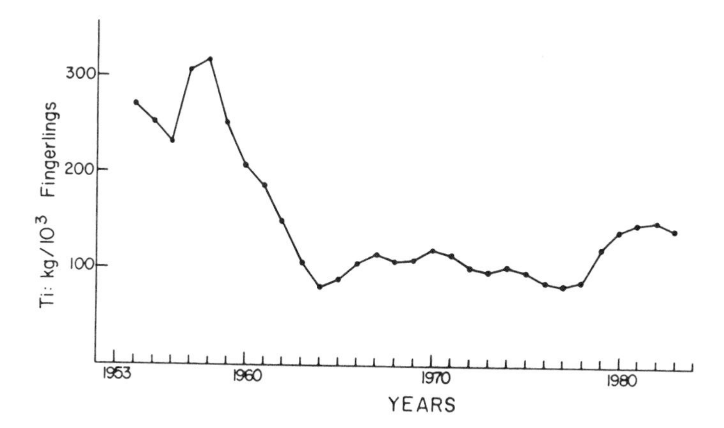

Figure 1. Running averages (3 consecutive years) of catch contribution (T_i) (kg catch /10^3 fingerlings/year) during 1953-1984 in Lake Kinneret. For T_i and I_i values, see text.

between catches and stocked fingerlings indicated best fit represented as $y = a_o X^{a1}$, and regressions were $\mathbf{y = 203.36\ X^{0.67}}$; $r^2 = 0.75$; $r = 0.87$, (Figure 2), where:

y = kg (catch)/10^3 introduced fingerlings/year.
X = introduced fingerlings in 10^6/year.

Results (Figure 1) show that the highest catch was achieved when annual introduction varied between 0.5 - 6.0 x 10^6 fingerlings/year. The contribution of S. galilaeus, O. aureus, silver carp and grey mullet to the catcher were 35%, 12%, 21% and 32% of stocked fish, respectively (Table 3).

TABLE 3 INCOME (10^6 \$) FROM STOCKED SPECIES IN LAKE KINNERET[a]

NAME	PRICE ($ per kg)	ANNUAL CATCH (tons)	INCOME (10^6 \$)	(%)
Grey mullet	4.0	200	0.8	40
St. Peter's fish	3.3	220	0.7	35
Silver carp	2.0	130	0.3	15
Blue tilapia	2.0	80	0.2	10

[a] The remainder of the total catch is 1170 tons, comprised of bleaks (990) and others, with a commercial value of \$1 x 10^6.

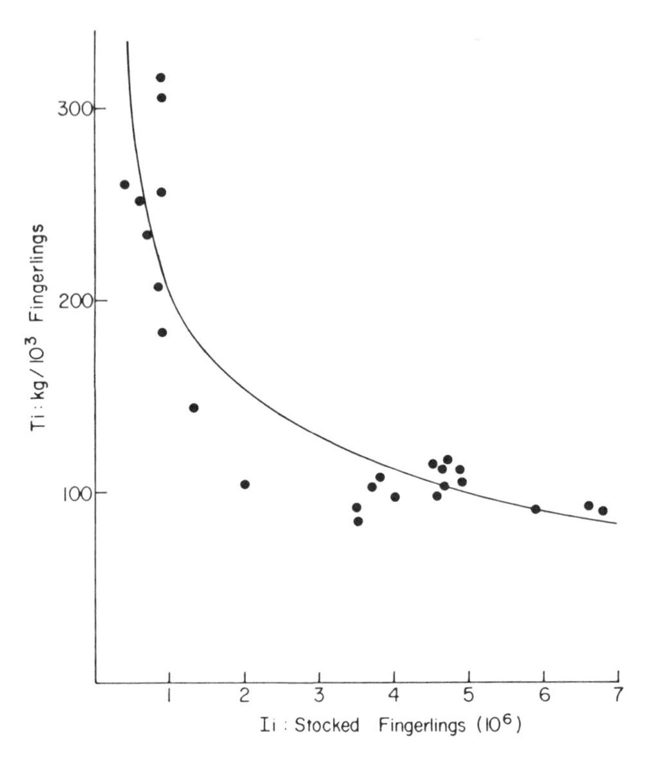

Figure 2. Best fit curve between catch contribution (T_i) and stocking (I_i) see text.

FISH FOOD SOURCES AND FOOD WEB STRUCTURE

Two food chains were found in the Kinneret ecosystem (Serruya et al. 1980), of which the summer structure includes nutrient limited nanoplankton (Serruya et al. 1980, Gophen and Pollinger 1985). In the winter season, there is a huge biomass of the dinoflagellate Peridinium cinctum in the lake, which is available as a food source for fish. However, only S. galilaeus is known as an efficient utilizer of this source (Spataru 1976, Serruya et al. 1980). About 10% of the Peridinium biomass is removed from the lake by the NWC and local pumping systems. Peridinium is easily removed from supplied water by low chlorination which damages the flagellae, causing the heavy cells to settle rapidly. The remainder of the Peridinium biomass is partly consumed by S. galilaeus (approximately 3- 7 % at present), but most if it is broken down in the water column and contributes organic substances to the summer assemblages (Serruya et al. 1980).

Lake conditions in summer dramatically change (Serruya et al. 1980, Gophen and Pollingher 1985, Gophen 1986). The summer ecosystem is nutrient limited and recycling processes are prominent. The dominant nanoplanktonic algae are primarily affected by nutrient levels and secondarily controlled by zooplankton grazing activity. Nevertheless, zooplankton biomass in the summer ecosystem of Lake Kinneret is efficiently controlled by fish predation. Food requirements of fish in summer are high because of the increased temperature (Gophen 1986), and the most available food source for fish in summer is zooplankton. Results in Table 4 show that 10-100% of the gut contents of commercial fish is zooplankton. Fish stock biomass in Lake Kinneret varies between 6000-7000 tons, and the biomass of commercial fishes comprised about 70% of the total (Serruya et al. 1980, Gophen 1986). Consequently, predation pressure of fish on zooplankton in summer is

TABLE 4 PERCENTAGE OF ZOOPLANKTON BIOMASS IN GUT CONTENTS OF MAJOR KINNERET FISH SPECIES (Spataru 1976, Spataru and Zorn 1978, Gophen 1980, Drenner et al. 1982b, Spataru and Gophen 1985 a,b,c, Spataru and Gophen unpublished data)

NAME	SUMMER-FALL	WINTER-SPRING
Sarotherodon galilaeus	10	6
Oreochromis aureus	35	41
Silver carp	59	9
Tristramella sacra	24	19
Tristramella simonis	32	11
Tilapia zillii	42	42
Mirogrex sp.:		
Adults	90-100	90-100
Fingerlings	80- 90	80-90
Fingerlings of cichlids and *Clarias*	70- 90	70-90
Clarias gariepinus	negligible	
Capoeta damascina	4	4
Astatilapia flaviijosephi	negligible	
Barbus longiceps	1	1
Tor canis	2	2

very intense. Moreover, intensifying fish predation pressure, mainly in summer (Landau 1980-1985, Gophen and Pollingher 1985), reduced zooplankton biomass, which resulted in an increase of nanoplankton densities (Figure 3). The introduction of fingerlings of O. aureus, silver carp and grey mullet, and increasing the bleak population in Lake Kinneret (Landau 1980-1985) enhanced predation pressure of fish on zooplankton (Gophen and Pollingher 1985).

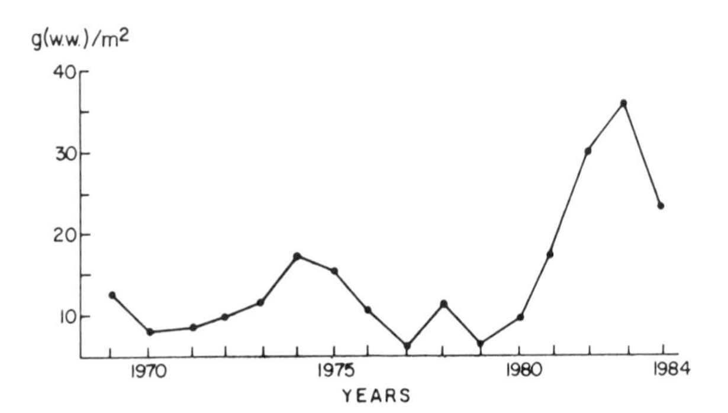

Figure 3. Non-Peridinium algal biomass (g (w.w.)/m^2) in Lake Kinneret, as yearly averages during 1969-1984 (Pollingher 1980-1985).

CHANGES IN ZOOPLANKTON COMMUNITIES

Over the years, the total zooplankton biomass in Lake Kinneret has shown a gradual reduction with regression values of $a_1 = -1.5\ g_{(w.w.)}/m^2/year$ ($P < 0.01$) and $r^2 = 0.79$ (Gophen 1985a). Additional evidence for intensifying fish predation is the extinction of Daphnia lumholtzi from the lake. This large zooplankter was commonly distributed until the late 1950s, and Lake Kinneret was the northernmost zoogeographical limit of this species. Some time between 1957 and 1963, D. lumholtzi disappeared. It is suggested that stocked fingerlings of grey mullet and O. aureus caused the extinction of this Daphnia species (Gophen 1979).

Results in Figure 4 indicate that there is a multinannual increase of the small size class abundance of cladocerans expressed as ratios of small/large organisms (Gophen 1985c). The reduction in body size of invertebrate grazers, as affected by intensification

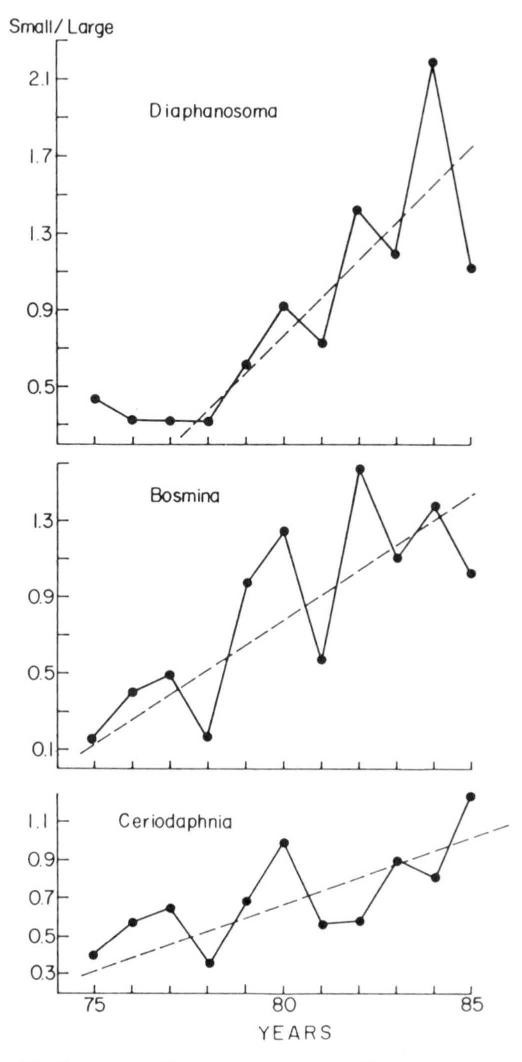

Figure 4. Small/large cladoceran ratios (Ceriodaphnia spp., Bosmina spp. and Diaphanosoma sp.) in Lake Kinneret during 1975-1985, presented as yearly averages.

of fish predation pressure, was shown by many authors (Hrbacek et al. 1961, Brooks and Dodson 1965, Hall et al. 1976). This author suggests that the observed increase in the abundance of small cladocerans relative to large ones was enhanced by intensification of the fish predation pressure in Lake Kinneret.

In our studies the body size dimensions of Bosmina were measured. Bosmina sp. was chosen because of the good preservation conditions of this organism in old samples. Fifty animals, arbitrarily chosen from samples collected in three months (February, September and November) during 1969, 1972, 1975, 1979, 1980 and 1985 at the same station and depths in Lake Kinneret were measured. Two dimensions were recorded: carapace length (= maximal length excluding the head) and maximal width. Results are presented in Table 5 and Figure 5. Linear regression between body dimensions of all months and years indicated $r^2 = 0.91$. It is concluded that the two parameters (= body dimensions) similarly changed. Results in Figure 5 indicated body size reduction. Carapace lengths and widths of Bosmina show similar values. Thus, the monthly averages of these two dimensions of each year were t-tested against 1969 values (Table 5).

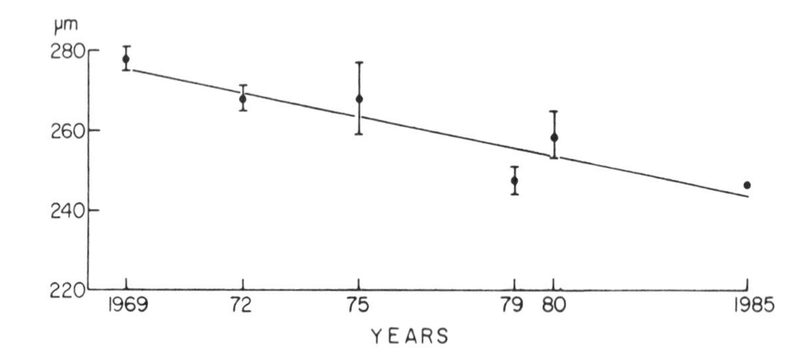

Figure 5. Yearly averages of carapace lengths and widths of Bosmina sampled in Lake Kinneret during three months (Feb., Sept. and Nov.) in 1969, 1972, 1975, 1979, 1980 and 1985.

Results in Table 5 show that Bosmina in Lake Kinneret became significantly smaller from 1979 throughout 1985. The "t" and "P" values (Table 5) indicate probabilities of 63-89% that Bosmina collected in Lake Kinneret during 1979, 1980 and 1985 were smaller than those samples in 1969. This is similar to data presented in Figure 4, although clear increases in the small/large ratio of cladocerans started earlier, in 1978.

TABLE 5 CARAPACE LENGTHS AND WIDTHS (um) OF BOSMINA: yearly averages of monthly means, t and p values of t-tested values for 1972, 1975, 1979, 1980 and 1985 against 1969

	YEAR					
	1969	1972	1975	1979	1980	1985
Yearly ave.	278	268	268	248	259	247
t values		0.517	0.429	2.080	1.054	2.182
p values		0.50	>0.50	0.11	0.37	0.12

FISH POISONING

In recent years, fisheries efforts on cichlids have been intensified accompanied by the lowering of the catch-per-unit-effort (Landau 1979). Recently, a fishing method consisting of using pesticides in the water has been observed (Wynne 1986). This is mostly carried out in the spawning ground area of the Beteicha lagoons (Gophen 1985) during the reproduction seasons of cichlids. The native impact of this activity on water quality, fish populations and public health is clear.

STOCKING AND FISHERIES MANAGEMENT: DISCUSSION AND RECOMMENDATIONS

It was documented that in spite of environmental improvements in the watershed area, and reduction of organic nitrogen in the lake (Gophen 1985b), water quality has slightly deteriorated (Pollingher 1980-1985, Gophen et al. 1983 b,c, Gophen 1985 b,c, 1986, Gophen and Pollingher 1985). This author suggests that this loss of quality is mostly due to food chain alterations by fish planktivory. To solve this problem, analyses of both economic and ecological requirements are needed. This is because the lake is primarily utilized for drinking and secondarily for fish exploitation.

Results presented here, and others (Pisanty, Ben-Yami and Talpax, unpubl. data) suggest that fishery exploitation in Lake Kinneret is maximal. Moreover, this author suggests that the introduction of more than 6×10^6 fingerlings a year would not contribute sufficient income as catches. Therefore the recommended introduction is for only those species that will give both water quality and economic benefits. The food consumption of grey mullet is, as yet, unknown, but its commercial value and income is the highest among the stocked species. The ecological benefit of S. galilaeus for water quality improvements has been previously discussed (Spataru 1976, Serruya et al. 1980, Gophen et al. 1983b,c, Gophen 1986 and others). This fish selectively and efficiently consumes Peridinium in winter and only slightly preys on zooplankton in summer (Table 4). The economic benefit of St. Peter's fish is presented in Table 3. The disadvantages of silver carp and blue tilapia refer to both water quality and fishermen's income. These fish are known as ineffective Peridinium consumers (Gophen 1980, Spataru and Gophen 1985c), since digestibles of the cells in fish guts are low compared to S. galilaeus. On the other hand, they intensively prey on zooplankton in summer (Table 4). The income from these two fishes is the lowest among stocked fish (Table 3), and they are efficiently cultured in fish ponds in Israel.

Seasonal high catches of silver carp and blue tilapia appear in the same months (except June catches of silver carp) as S. galilaeus and grey mullet (Table 6). Total stocked fish catches are 435 tons and 214 tons in the winter-fall (January-May, November-December) and summer (June-October) months, respectively (Table 6). Moreover, similar to S. galilaeus (73% in winter and 24% in summer catches), 64% and 35% of other stocked species are due to winter and summer month landings, respectively (Table 6). Thus silver carp and O. aureus do not compensate for low seasonal fisheries of native species (Golani 1980-1984). The recommended stocking program is based on the principle that total numbers would not be more than about 6×10^6/year.

Nevertheless, in order to increase grazing pressure of S. galilaeus on Peridinium, intensification of the stocking activity of this fish is required. Zooplankton resources in summer are limited; thus the elimination of other stocked species is needed. Due to low income and ecological impacts, the author recommends the elimination of silver carp and O. aureus from the stocking program. Stocking programs should include annually 1×10^6 and $4\text{-}5 \times 10^6$ fingerlings of grey mullet and St. Peter's fish, respectively. Because bleaks are purely zooplanktivorous as fingerlings and adults (Gophen and Landau 1977, Gophen and Scarf 1981), and their stock biomass was recently increased (Landau 1980 - 1984) the author suggests an intensification of their fisheries. To carry this out properly, their market demands should be improved.

TABLE 6 MONTHLY AVERAGES (S.D.) (1981-1984; no published data of monthly catches prior to 1981) of S. galilaeus, blue tilapia, silver carp and grey mullet catches, in %, in Lake Kinneret. Annual and seasonal (a: January-May and November-December; b: June-October) mean catches (monthly: X - %; T - tonnes) are given (Golani 1980-1984)

MO.	S. galilaeus	B. tilapia	S. carp	G. mullet
1	12.4 (5.0)	13.1 (8.9)	6.6 (4.8)	10.2 (6.4)
2	11.8 (4.5)	10.3 (4.5)	7.0 (3.8)	7.9 (1.0)
3	10.8 (3.5)	7.2 (7.3)	6.4 (3.2)	7.9 (3.0)
4	7.5 (3.8)	8.2 (6.1)	10.1 (4.3)	6.2 (1.6)
5	6.6 (4.6)	7.8 (4.8)	11.9 (1.3)	6.6 (2.4)
6	2.4 (1.0)	5.1 (1.9)	22.6 (8.3)	5.4 (1.9)
7.	4.2 (1.0)	6.3 (1.5)	8.7 (8.2)	6.8 (4.8)
8.	6.1 (2.3)	4.7 (0.4)	3.2 (1.9)	5.7 (5.9)
9.	7.0 (0.9)	6.1 (2.5)	4.3 (2.1)	7.2 (0.8)
10.	7.6 (1.3)	6.7 (2.3)	5.4 (2.6)	8.0 (2.8)
11.	8.6 (4.8)	14.0 (12.9)	7.3 (5.6)	13.6 (3.7)
12.	15.0 (11.3)	10.5 (8.0)	6.5 (4.4)	14.6 (7.4)
X:a	10.4 (3.0)	10.2 (2.6)	8.0 (2.2)	9.6 (3.4)
b	5.5 (2.1)	5.8 (0.8)	8.8 (8.0)	6.6 (1.1)
T:a	22.7	6.8	11.2	21.5
b	12.0	3.9	12.3	14.8
TOTAL				
T:a	158.9	47.4	78.4	150.2
b	59.4	19.5	61.2	73.6

CONCLUSIONS

Discussions between the limnological team of the Kinneret Institute and fishery biologists of the Ministry of Agriculture initiated a compromise new program of fisheries and stocking managements, started in 1984. The program will run for a three-year test period and is aimed at the prevention of both water quality deterioration and the reduction of quantity and quality of commercial fisheries to ensure reasonable incomes to the fishermen. The programs's goals will be achieved by the following operations:

1. Sarotherodon galilaeus (St. Peter's fish) - Increasing stocking of this fish in Lake Kinneret to an annual introduction of 4-5 x 10^6 fingerlings (approximate weight 6-10

g/fingerling) during March-April, and improving natural reproductive conditions (see part 5).

2.Silver carp - elimination of stocking.

3.<u>Oreochromis aureus</u> (blue tilapia) - Introduction of not more than 0.5×10^6 fingerlings a year.

4.Bleak fisheries - Intensifying exploitation of bleaks in the lake. This will be carried out by searching for new commercial bleak products or recipies. The Fisheries Department recommended enquiring about potential predators to be stocked in the lake after an experimental study of their predation behaviour and prey selectivities, but this was strongly objected to by this author.

5.Establishing a law preventing fishing activity in the shallow Beteicha lagoons in the northeastern part of the lake during the reproduction period of cichlids (April-June). This restricted area is the one remaining spawning ground for cichlids in the lake undisturbed by human recreational activities.

6.Prevention of the criminal use of pesticides for "fishing" purposes. Prevention of fishing activity in the lagoons of Beteicha will help police stop the criminal poisoning which is mostly done in this area during April-June.

7.Increasing legal mesh sizes of nets (especially purse seine) to ensure that only St. Peter's fish of legal commercial size (> 20 cm TL) are caught.

ACKNOWLEDGEMENTS

I warmly thank Dr.David Wynne for criticizing the manuscript and B. Azoulay for her technical assistance.

REFERENCES

Barel, C.D.N., Doris, R., Greenwood, P.H., Fryer, G., Hughes, N., Jackson, P.B.N., Kawanabe, H., Lowe-McConnell, R.H., Nagoshi, M., Ribbink, A.J., Trewavas, E., Witte, F. and Yamaoka, K. 1985. Destruction of fisheries in Africa's lakes. Nature 315:19-20.

Bartell, S.M. and Kitchell, J.F. 1978. Seasonal impact of planktivory on phosphorus release by Lake Wingra zooplankton. Verh. Int. Ver. Limnol. 20:466-474.

Ben-Tuvia, A., Reich, K. and Ben-Tuvia, S. 1983. Stocking and management of cichlids fishery resources in Lake Kinneret (Lake Tiberias). In: Proceedings of the Int. Symp. on Tilapia in Aquaculture. pp. 18-27.

Brookes, J.L. and Dodson, S.L. 1965. Predation body size and composition of plankton. Science 150:28-35.

Carpenter, S.R., Kitchell, J.F. and Hodgson, J.R. 1985. Cascading trophic interactions and lake productivity. Biosci. 35:634-639.

Drenner, R.W., Denoyelles, F. and Kettle, D. 1982a. Selective impact of filter-feeding gizzard shad on zooplankton community structure. Limnol. Oceanogr. 27:965-968.

Drenner, R.W., Vinyard, G.L., Gophen, M. and McComas, S.R. 1982b. Feeding behaviour of the cichlid Sarotherodon galilaeus: selective predation on Lake Kinneret zooplankton. Hydrobiologia 87:17-20.

Fernando, C.H. and Holcik, J. 1982. The nature of fish communities. A factor influencing fishery potential and yields in tropical lakes and reservoirs. Hydrobiologia 97:127-140.

Golani, D. 1980-1984. Lake Kinneret. In: Israel Fisheries in Figures. pp. 20-24. Israel: Minist. Agric. Dep. Fish.

Gophen, M. 1979. Extinction of Daphnia lumholtzi (Sars) in Lake Kinneret (Israel). Aquaculture 16:67-71.

Gophen, M. 1980. Food sources, feeding behaviour and growth rates of Sarotherodon galilaeum (Linnaeus) fingerlings. Aquaculture 20:101-115.

Gophen, M. 1984. The Impact of zooplankton status on the management of Lake Kinneret (Israel). Hydrobiologia 113:249-258.

Gophen, M. 1985a. Zooplankton. In: Lake Kinneret. Annual Report. ed. M. Gophen, pp. 82-83. Tabgha: Yigal Allon Kinneret Limnol. Lab.

Gophen, M. 1985b. The management of Lake Kinneret and its drainage basin. In: Scientific Basis for Water Resources Management. Proceedings of the Jerusalem Symposium, IAHS Publ. No. 153, pp. 127-138.

Gophen, M. 1985c. Effect of fish predation on size class distribution of cladocerans in Lake Kinneret. Verh. Int. Ver. Limnol. 22:3104-3108.

Gophen, M. 1986. Fisheries management in Lake Kinneret (Israel). In: Proceedings of the Ann. Mtg., N. Amer. Lake Mgmt. Soc. (in press).

Gophen, M., Drenner, R.W. and Vinyard, G.L. 1983a. Fish introduction into Lake Kinneret: call for concern. Fish. Mgmt. 14:43-45.

Gophen, M., Drenner, R.W. and Vinyard, G.L. 1983b. Cichlid stocking and the decline of the Galilee St. Peter's fish (Sarotherodon galilaeus) in Lake Kinneret, Israel. Can. J. Fish. Aquat. Sci. 40:983-986.

Gophen, M., Drenner, R.W., Vinyard, G.L. and Spataru, P. 1983c. The cichlid fishery of Lake Kinneret: history and management recommendations. In: Proceedings of the Int. Symp. on Tilapia in Aquaculture. pp. 1-7.

Gophen, M. and Landau, R. 1977. Trophic interactions between zooplankton and sardine Mirogrex terraesanctae populations in Lake Kinneret, Israel. Oikos 29;166-174.

Gophen, M. and Pollingher, U. 1985. Relationships between food availability, fish predation and the abundance of the herbivorous zooplankton community in Lake Kinneret. Arch. Hydrobiol. Beih. Ergebn. Limnol. 21:397-405.

Gophen, M. and Scharf, A. 1981. Food and feeding habits of Mirogrex fingerlings in Lake Kinneret (Israel). Hydrobiologia 78:3-9.

Hall, D.J., Threlkeld, S.T., Burns, C. and Crowley, P.H. 1976. The size efficiency hypothesis and the size structure of zooplankton communities. Ann. Rev. Ecol. Syst. 7:177-208.

Hrbacek, J., Dvorakova, M., Korinek, V. and Prochazkova, L. 1961. Demonstration of the effect of the fish stock on the species composition of zooplankton and the intensity of metabolism of the whole plankton assemblage. Verh. Int. Ver. Limnol. 14:192-195.

Kitchell, J.F. 1980. Fish dynamics and phosphorus cycling in lakes. In: Nutrient Cycling in the Great Lakes: A Summarization of the Factors Regulating Cycling of Phosphorus. ed. D. Scavazi and R. Moll, pp. 81-91. Ann Arbor, MI: Great Lakes Environ. Res. lab., NOAA Spec. Rep. 83.

Landau, R. 1979. Growth and population studies on Tilapia galilaea in Lake Kinneret. Freshwat. Biol. 9:23-32.

Landau, R. 1980-1985. Fish population dynamics. In: Lake Kinneret. Annual Report. ed. M. Gophen, pp. 20-23. Tabgha: Yigal Allon Kinneret Limnol. Lab.

Larkin, P.A. 1978. Fish management - an assay for ecologists. Ann. Rev. Ecol. Syst. 9:57-74.

Lehman, J.T. 1980. Release and cycling of nutrients between planktonic algae and herbivores. Limnol. Oceanogr. 25:620-632.

Pollingher, U. 1980-1985. Phytoplankton. In: Lake Kinneret. Annual Report. ed. M. Gophen, pp. 25-28. Tabgha: Yigal Allon Kinneret Limnol. Lab.

Rigler, F.H. 1982. The relation between fisheries management and limnology. Trans. Am. Fish. Soc.111:121-132

Sarid, Z. 1950-1979. Lake Kinneret. In: Israel Fisheries in Figures. Israel: Minist. Agric. Dep. Fish.

Serruya, C., Gophen, M. and Pollingher, U. 1980. Lake Kinneret: carbon flow patterns and ecosystem management. Arch Hydrobiol. 88:265-302.

Serruya, Pollingher, U. Cavari, B.Z., Gophen, M., Landau, R. and Serruya, S. 1979. Lake Kinneret management options. Arch. Hydrobiol. Beih. Ergebn. Limnol. 13:306-316.

Shapiro, J. 1980. The importance of trophic level interactions to the abundance and species composition of algae in lakes. In: Hypertrophic Ecosystems. ed. J. Barica and L.R. Mur, pp. 105-115. The Hague: Junk.

Shapiro, J., Forsberg, B., Lamarra, V., Lindmark, G., Lynch, M., Smeltzer, E. and Zoto, G. 1982. Experiments and experience in biomanipulation. Studies of biological ways to reduced algal abundance and eliminate blue-greens. U.S. Environmental Protection Agency. Report number EPA 600/3-82-096, 235 pp.

Shapiro, J. and Wright, D.I. 1984. Lake restoration by biomanipulation. Freshwat. Biol. 14:371-383.

Spataru, P. 1976. The feeding habits of Tilapia galilea (Artedi) in Lake Kinneret (Israel). Aquaculture 9:47-59.

Spataru, P. and Gophen, M. 1985a. Food composition and feeding habits of Astatilapia flaviijosephi (Lortet) in Lake Kinneret (Israel). J. Fish. Biol. 26:503-507.

Spataru, P. and Gophen, M. 1985b. Food composition of the barbel Tor canis (Cyprinidae) and its role in the Lake Kinneret ecosystem. Environ. Biol. Fish. 14:295-301.

Spataru, P. and Gophen, M. 1985c. Feeding behaviour of silver carp Hypophthalmichthys molitrix Val. and its impact on the food web in Lake Kinneret, Israel. Hydrobiologia 120:53-61.

Spataru, P. and Zorn, M., 1978. Food and feeding habits of Tilapia aurea (Steindachner) (Cichlidae) in Lake Kinneret (Israel). Aquaculture 13:67-69.

Wynne, D. 1986. The potential impact of pesticides on the Kinneret and its watershed over the period (1980-1984). Environ. Pollut. Ser. A. (in press)

Zaret, T.M. and Paine, R.T. 1973. Species introduction in a tropical lake. Science 182:449-455.

CHAPTER 3

FISHERY TRANSFORMATION ON LAKE MANZALA, EGYPT DURING THE PERIOD 1920 TO 1980

Don R. Toews

Department of Natural Resources, Box 20, 1495 St. James Street, Winnipeg, Manitoba, R3H 0W9

ABSTRACT

Lake Manzala, a brackish lake in the Nile Delta region of Egypt has been impacted by anthropological and environmental factors over the past 60 years. During this period salinities have declined and nutrient loading has increased on Lake Manzala, in response to a two or three fold increase in agricultural drainwater flows, reduced evaporation due to approximately 50 percent reduction in lake area resulting from agricultural land reclamation and increased domestic sewage outfall from Cairo. Impacts of long-term water quality changes to fishing are: a transformation from a marine mullet to a freshwater tilapia-dominated fishery; a seven fold increase in yield per unit area; a two or three fold increase in gross catch value per unit area; a four fold increase in primary employment in terms of number of fisherman per unit area. Existing regional differences in primary production, fish standing stock, fish yield and fishing effort are a function of drainwater impact. Areas unaffected by drainwater have not significantly changed over time. Levels of heavy metals and pesticides in the water column in sediments and in fish are somewhat elevated in drainwater outfall areas but within acceptable levels. Nutrient loads are expected to double by the year 2000. It is proposed that long term fisheries development on Lake Manzala be based on the enriched freshwater tilapia model and on the optimal utilization of existing and projected future increases in nutrient loading.

INTRODUCTION

The lakes and rivers of the Nile Delta region in northern Egypt represent a dynamic system that has been undergoing continuous and pronounced changes since ancient times (Montasir 1937, ARE 1977). Natural and cultural influences on Lake Manzala, and the corresponding physical, chemical, hydrological and biological changes have been documented to some extent since the early 1900s. A review and evaluation of these data are keys to understanding the dynamics of the present fishery, to predicting fishery responses to further natural and man-induced environmental changes, and as a basis for management.

Annual government reports from the early part of this century provide fish catch, effort and salinity data for a period of relative fisheries and environmental stability (Paget 1924, Fouad 1928, Wimpenny 1930, Faouzi 1936, Samra 1933). Several extensive research studies in the areas of water quality, lake morphology, ecology and fisheries, were carried out during the 1060s and early 1970s (Wakeel and Wahby 1970a and 1970b, Wahby et al. 1972, Bishara 1973, Youssef 1973, Shaheen and Youssef 1978, 1979, Panse and Sastry 1960), prior to and during a period of major environmental changes on Lake Manzala. The results of the fishery and ecological field programs carried out during 1979 and early 1980, as part of the Lake Manzala Study (Maclaren 1981, Toews and Ishak 1984), may be regarded as a post-assessment of this period. This is not to suggest, however, that the lake and its associated fisheries have stabilized, as many of the influences, especially in cultural terms, are ongoing or projected to continue for the future.

This paper first provides an overview of the existing fisheries and environmental conditions on Lake Manzala. This is followed by a historical review of the fishery to identify changes and trends of key limnological and fisheries parameters. An attempt is made to relate past fishery states to present conditions in different areas of Lake Manzala and to evaluate the fishery transformations which have occurred. Based on this assessment some management and future development considerations are identified and discussed.

OVERVIEW OF THE LAKE MANZALA FISHERY

The Lake Manzala fishery is comprised of an open water area of 89,000 ha which sustains a conventional "hunt and capture" open access fishery yielding an estimate 41,000 tonnes with a first sale value of LE14.6 million (one Egyptian pound (LE) = approximately $1.50 U.S.). The open water fishery employs approximately 17,000 fishermen on a full time basis of which an estimated 4,000 are pre-adolescent boys.

In addition to the open water fishery, two general types of fish culture (closed fishing) operations are found within and around Lake Manzala - lake based hosha and land based fish farms. Hosha involve the enclosure of open water areas by impermeable earthen dykes with several openings connected to the open lake. Uncontrolled stocking occurs from adjacent open water areas. Periodically the openings are plugged, the hosha pumped dry and the fish harvested. An estimated 15,000 ha of hosha located on the main body of Lake Manzala are closely linked with open water fisheries in terms of water quality and fish stocks. Hosha function in part as a harvesting mechanism, and a major portion of the hosha yield is derived from open lake production. A further 4,600 ha of lake based marine fish culture operations in the northern sector are considered as fish farms. In general these operations are closed systems and interrelationships with open waterstocks are minimal and need not be considered in the calculation of open water yields. Extensive land based fish farming operations involving controlled flooding and stocking of dry lands are found to the south and southeast of the lake along agricultural drains.

For the purposes of this study the open water area of Lake Manzala has been divided into four generalized areas or sectors distinguished in terms of water quality, nutrient loading, fisheries effort and productivity (Figure 1). A fifth area, Lake Um El Rish is geographically separated from the main lake proper.

HYDROLOGY, WATER QUALITY AND PRODUCTIVITY

The southern sector of Lake Manala serves as a receiving basin for the three major agricultural drains (Figure 1) which account for about 78 percent of the fresh water inflow into the lake and 84 percent of the total annual nutrient loading (total nitrogen plus

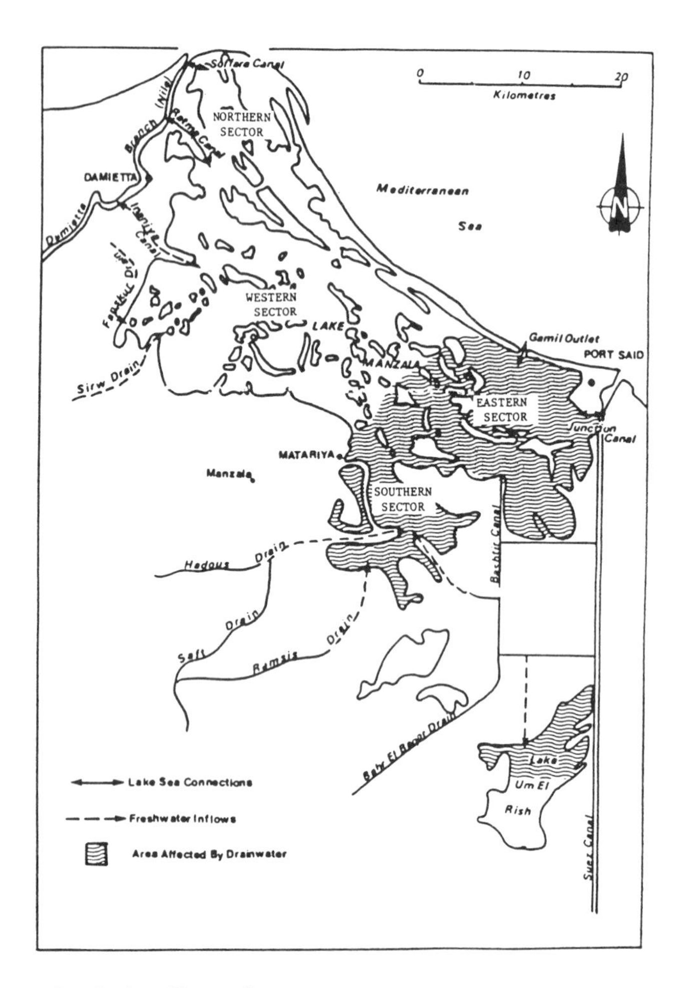

Figure 1. Lake Manzala.

total phosphorus) estimated at 13,000 tonnes (Table 1). Cairo sewage discharged via the Bahr El Baqar is the major source of nutrients for the lake. It accounts for 90 percent of the nutrients transported by the drain and 58 percent of the total existing nutrient load being discharged into the lake. The major outflow from the lake is at the northeast end of the lake via the Gamil outlet and Junction Canal which account for 80 percent and 17 percent respectively of the total outflow. Because of the inflow and outflow configuration and prevailing wind direction the lake is

TABLE 1 DRAINWATER DISCHARGE AND NUTRIENT LOADING INTO LAKE MANZALA DURING THE PERIOD 1979-1980 (MacLaren 1981, volume 6).

Drainwater Source	Discharged Flow ($m^3 x 10^6$/yr)	Percent of Total Flow (%)	Nitrogen Concentration (mg/L-N)	Nitrogen Load (t/yr-N)	Phosphorus Concentration (mg/L-P)	Phosphorus Load (t/yr-P)	Nutrient Load (t/yr)	Percent of Total Nutrient Load (%)
Bahr El Baqar Drain	1 678	25	3.683	6 180	1.013	1 700	7 880	60
Hadous Drain	3 276	49	0.730	2 391	0.192	629	3 020	23
Ramsis Drain	252	4	0.633	160	0.143	36	196	1
Sirw Pumpstation	847	13	1.144	969	0.181	153	1 122	8
Matariya Pumpstation	154	2	0.381	59	0.18	28	87	1
Fariskur Pumpstation	292	4	0.450	131	0.138	40	171	1
Port Said Sewage	40	1	14.888	596	4.128	165	761	6
Inaniya Canal	156	2		Nutrient load is small and has been neglected				
Total	6 695	100	-	10 486	-	2 751	13 237	100

not well mixed and the bulk of the nutrient rich drainwater entering the southern sector flows northeast to the two major outlets heavily impacting the eastern sector in terms of nutrient loading, salinity and productivity. The western sector receives relatively low levels of freshwater inputs and nutrient loading via the Sirw and Fariskur pumping stations and the Inaniya Canal. The northern sector receives no local fresh water and exhibits the highest salinity levels in the lake. Salinities are further elevated by salt water intrusions from the Mediterranean via the Ratma and Soffara Canals to compensate for evaporation losses during the summer period. Nutrient rich water from the Bahr El Baqar Drain was diverted into the northern end of shallow Lake Um El Rish in 1978. During the summer the lake acts as an evaporation basin resulting in a continuous inflow of drainwater and extensive nutrient loading estimated at 700 tonnes per annum. This has resulted in relatively low salinity and a highly productive tilapia fishery at the northern end of Lake Um El Rish. Fish production at the southern end of the lake is depressed by high salinity.

Seasonal mean salinities range from around 2,000 mg/L in areas impacted by drainwater to 15,000 mg/L in the coastal areas (Table 2). Salinities of over 40,000 mg/L were reported in shallow protected areas of the northern sector during high evaporation periods.

Primary production levels and fish standing stocks generally reflect the level of nutrient loading by lake sector (Table 2). The saline northern sector which has negligible levels of nutrient loading exhibited the lowest primary production levels of 2-3 g O_2/m^2.day and fish standing stocks of less than 250 kg/ha. The western sector which has low levels of nutrient loadings had primary production levels of 3-5 g O_2/m^2.day and fish standing stocks of 250-500 kg/ha. The highest levels of primary production were in the eastern and southern sectors which are heavily impacted by drainwater. Recorded values ranged from 6-23 g O_2/m^2.day. Fish standing stocks were 500 - 1,000 kg/ha and 1,700-3,000 kg/ha in the eastern and southern sector, respectively. Standing stock levels in the northern drainwater affected area of Lake Um El Rish were in the 500-1,000 kg/ha category while primary production based on one measurement was 5 g O_2/m^2.day.

TABLE 2 SUMMARY OF SALINITY, NUTRIENT LOADING, PRIMARY PRODUCTION AND FISH STANDING STOCKS BY LAKE SECTOR FOR LAKE MANZALA. (for detailed data analysis and methodology see MacLaren 1981, volumes 7,8 and 10).

Lake Sector	Salinity (mg/L		Primary Productivity ($gmO_2/m^2/day$) Range	Nutrient Loading	Fish Standing Stocks (kg/ha)
	Mean	Range			
Northern	14 800	2 500-36 000	2-3	Very low	0- 250
Western	2 100	400- 5 000	3-5	Low	250- 500
Eastern	3 500	1 250-22 000	6-23	High	500-1 000
Southern	1 700	1 000- 4 000	8-19	Very high	1 700-3 000
Lake Um El Rish			5	High localized	500-1 000

CATCH AND EFFORT

Twenty-seven species (24 fish and 3 crustacean) have been identified in the Lake Manzala open fishery catch (Table 3). The four tilapia species account for about 35,000 tonnes, or 85 percent of the total annual catch. The species Sarotherodon nilotica represents 63 percent of the tilapia catch.

TABLE 3 SPECIES COMPOSITION OF THE OPEN WATER FISH CATCH ON LAKE MANZALA DURING MAY 1979 TO APRIL 1980 (MacLaren 1981, volume 10).

		Volume (t)	% of Total	% of Species
Tilapia[a]	Sarotherodon nilotica	21 927	54	(63)
	S. aurea	7 259	18	(21)
	S. galilaea	2 312	6	(7)
	Tilapia zillii	3 325	8	(10)
	Total:	34 823	86	(100)
Mullet	Liza ramada	776	2	(86)
	Mugil cephalus	127	<1	(14)
	Total:	903	2	(100)
Catfish	Bagrus bayad	1 103	3	(58)
	Clarias lazera	812	2	(42)
	Total:	1 915	5	(100)
Eels	Anguilla anguilla	309	1	
Other marine	Dicentrarchus labrax	227	1	(89)
	D. punctata	9	-	(4)
	Sparus auratus	1	-	<1
	Hemiramphus sp.	17	-	(6)
	Miscellaneous[c]	48	<1	
	Total:	302	1	(100)
Other fresh-water	Lates niloticus	4	-	(5)
	Labeo niloticus	22	<1	(27)
	Cyprinus carpio	6	-	(7)
	Dalophis imberbis	49	<1	(61)
	Total:	81	<1	(100)
Shrimp	Palaeomon elegans[b]	2 312	6	(95)
	Metapenaeus stebbingi	118	<1	(5)
	Total:	2 430	6	(100)
TOTAL		40 761	100	

[a]Includes less than one percent Haplochromis desfontainesii and Hemichromis bimaculatus.

[b]Includes less than one percent mullet fry.

[c]A mixture of small D. labax, D. punctata, S. auratus, Hemiramphus sp. and Solea solea.

A total of 900 tonnes of mullet were caught, amounting to two percent of the total catch. Of this, 86 percent were the species *Liza ramada*.

The highest levels of yield and fishing effort are found in the southern sector of Lake Manzala (Table 4). Over 50 percent or an estimated 21,000 tonnes of the total annual open fishery yield of 41,000 tonnes is derived from this region. This represents an impressive annual area yield of 2,016 kg/ha. Virtually 100 percent of this is low value species with tilapia comprising 92 percent of the catch. Effort is high with an estimated 7,200 fishermen and 1,200 boats operating in the region. This represents a density of 69 fishermen and 11.4 boats per 100 ha of lake surface. In spite of intensive effort the CPUE (catch per unit effort) is still the highest on the lake at 3.4 tonnes/fisherman with the exception of Lake Um El Rish which was a developing fishery experiencing a fishing-up process. The southern sector is characterized by a high proportion of large boats and well capitalized fishing operations. Average crew size of 6.1 fishermen per boat is the highest on the lake. Gross and net returns per hectare are high at LE 808 and LE 574, respectively. The high biological productivity of the southern sector can largely be attributed to the inflow of nutrient-rich water from the Bahr El Baqar Drain. Mullet are almost completely absent from the area, probably because of eutrophic conditions. Salinity does not appear to be a limiting factor since mullet are found in other areas with similar low salinities, e.g. the western and eastern sectors.

The eastern sector accounts for 13,200 tonnes or 32 percent of the total lake catch. This region is heavily influenced by drainwater from the southern sector. While average salinity levels in the area are low, regional and seasonal elevations occasionally occur due to saltwater inflow via the two lake-sea connections at Gamil and Junction Canals. Yields are moderately high averaging 435 kg/ha. A significant proportion of the catch (8 percent) is made up of high value marine species which enter the lake via the two lake-sea connections. During summer and autumn, effort in the area increases as the fishery concentrates on emigrating marine species, especially mullet. At this time of year the marine component of the catch increased to over 30 percent. The relatively high fishing effort of 21.9 fishermen per 100 hectares and low CPUE of 1.5 tonnes per fisherman probably reflects this concentration of effort on high value marine species. Gross and net returns are at an intermediate level of LE 203 and LE 122 per ha respectively.

TABLE 4 ESTIMATES OF OPEN FISHING OUTPUT AND EFFORT BY LAKE SECTOR (for more detailed data see MacLaren 1981, volume 10).

	Lake Sector				
	Northern	Western	Eastern	Southern	Um El Rish
Area (ha)	9 760	30 930	30 480	10 500	7 800 (2 340)
Annual Catch (t)	822	3 819	13 245	21 160	2 267
Fish Yields (kg/ha)					
Mullet	6	15	14	1	<1 (1)
Crustaceans (high value)	-	-	4	-	-
Other Marine	1	1	18	1	-
Total High Value	7	16	36	2	<1 (1)
Tilapia	77	103	311	1 866	286 (952)
Crustaceans (low value)	-	-	63	36	-
Other Fresh Water	<1	5	24	111	5 (16)
Total Low Value	77	108	298	2 013	291 (968)
Total Yield (kg/ha)	84	124	435	2 016	291 (969)
Gross Return (LE/ha)	39	62	203	808	116 (389)
Net Return (LE/ha)	27	42	122	574	97 (324)
Fishing Effort					
Total Number of Fishermen	409	1 784	6 681	7 259	459
Total Number of Boats	215	625	1 725	1 190	255
Number Fishermen/100 ha	4.3	5.7	21.9	69.0	6.0 (19.5)
Number Boats/100 ha	2.1	1.9	5.7	11.4	3.3 (11.0)
Number Fishermen/Boat	1.9	2.9	3.9	6.1	1.8
Catch Per Unit Effort (CPUE)					
t/Fisherman	2.1	1.7	1.5	3.4	4.9
t/Boat	3.8	6.1	7.7	17.8	8.9

[a]Adjusted value takes into account the restriction of fishing activity to only 30 percent of Lake Um El Rish.

[b]Values based on economic prices. Main shadow price assumptions are High value fish LE1.20/kg, Low value fish LE0.40/kg and unskilled labour LE1.0/man/day (MacLaren, 1981, Vol. 6).

Estimated yields in the western sector are relatively low at 3,800 tonnes or 124 kg/ha. The small annual catch per fisherman (1.7 tonnes) coupled with low fishing effort (5.7 fishermen/100 ha) is indicative of low potential yield for the region. This sector is of particular importance as a mullet rearing area and provides approximately 50 percent of the total open fishery mullet output of around 900 tonnes. Despite relatively low salinity (averaging 2,000 mg/L and the long distance from the lake-sea connection, mullet comprise 12 percent of the total recorded catch, the highest proportion of any lake sector.

The northern sector can be regarded as a low intensity, low yield fishing area. Fishing units are small scale operations. Almost 70 percent use trammel nets and the remainder use either wire baskets or spiral shaped set traps. Mean crew size is small (1.9 fishermen/boat) and fishing effort in the open water is low (4.3 fishermen/100 ha). Salinities in this open water are very high and over 90 percent of the open water commercial catch is comprised of a single species, the salinity tolerant *Tilapia zillii*.

Lake Um El Rish was connected to Lake Manzala via branches of the Bahr El Baqar Drain shortly before the 1979-1980 study period. The nutrient rich flow from the drain created a situation similar to that in the southern sector. However, the nutrient flow and subsequently the fishing activity are restricted to the northern 30 percent of the lake area round the mouth of the drain. Based on this estimate of reduced fisheries area, current annual yields are second only to the southern sector at 969 kg/ha. The recorded annual yield of 2,300 tonnes is based on catch estimates of an expanding fishery and is probably an underestimate. The annual catch per fisherman of 4.9 tonnes is the highest on Lake Manzala and is likely indicative of a fishing-up phase. Based on catch results of the period August 1979 to April 1980 and seasonal yield patterns in other lake areas, the annual yield for the present fishery could be as high as 3,500 tonnes.

FISHING GEAR AND METHODS

Many different types of fishing gear and methods are employed on Lake Manzala. This diversity of effort, which exhibits distinct geographical and seasonal patterns, effectively exploits most fish stocks in the lake.

A detailed description of fishing methods is complicated by the fact that most methods, while used throughout the lake, are referred to by their local names, which change from region to region. For detailed treatment of this subject see Bishara 1973, Youssef 1973 and Shaheen and Youssef 1979.

Trammel nets are the dominant fishing gear on Lake Manzala, producing a total of 11,600 tonnes or 28 percent of the estimated annual catch. Roughly 50 percent of the fishing units or 2,000 boats utilize trammel gear and methods. Other major fishing methods used are seine and the seine/hand combination. Each accounts for about 8,000 tonnes or 21 percent of the annual catch. Other methods include wire traps, spiral fence traps, trawled frame nets and surround netting of floating vegetation mats. While trammel and seine methods are relatively non-selective for species, spiral traps are highly selective for mullet, eels and other marine species. Surround net methods select for catfish, especially Clarias lazera. Seine/hand combinations, frame nets and hand fishing methods catch almost exclusively tilapia. Hand and seine/hand combination methods catch primarily Sarotherodon nilotica.

AN HISTORICAL REVIEW

The area of Lake Manzala has been reduced from 171,000 ha during the early 1900s to about 147,000 ha in 1949 and to 126,000 ha during the 1960s. In the past decade, extensive agricultural reclamation activities on the southern and southwestern sides of the lake, the expansion of hosha and settlements on the islands have further reduced the open water area to 89,000 ha. The mean depth of Lake Manzala was about 1 m in 1979 while the maximum depth was 2 m.

WATER QUALITY AND HYDROLOGY

Lake Manzala has traditionally been a brackish (partly salt) lake because of its connections with the Mediterranean Sea. The lake has sustained a fish stock comprising a mixture of species ranging from fresh water (e.g. tilapia) to marine (e.g mullet). The success of the various species (i.e. their relative contribution to the fish catch over the years) has been influenced by salinity to some degree.

Annual mean salinity levels in Lake Manazal (northern sector excluded) declined by about 50 percent; from 16,700 mg/L during the period 1921 to 1926, to 9,000 mg/L in 1962 to 1963. During the rest of the 1960s salinity levels remained relatively constant, but dropped to 4,100 mg/L in 1973 and to 2,900 mg/L during 1979 to 1980. The salinity of Lake Manzala is a function of the hydrology of the system (i.e. the balance and quality of inflow and outflows). Inflows are comprised of fresh water from the agricultural drains on the south side of the lake: Nile water, via the Inaniya Canal, on the western side; salt water on the northern side, through lake-sea connections (Figure 1). Outflows occur through lake-sea connections, and by evaporation from the lake surface. The effect of evaporation on salinity in Lake Manzala is pronounced because of its shallow mean depth. Since the 1930s there has been a noticeable decrease in evaporation from the lake due to the lessening of the lake area. This, coupled with an increase of fresh water inflow, has reduced levels of salinity. The decline in relative importance of evaporation as a factor in lake hydrology (and salinity) is demonstrated in Table 5.

TABLE 5 SUMMARY OF LAKE HYDROLOGY DURING 1933 TO 1978 (MacLaren 1981, volume 10).

Parameter	Period 1933-35	1962-68	1974-78
Fresh water inflow ($m^3 x 10^6$)	1 209	5 864	6 656
Rainfall ($m^3 x 10^6$)	190	140	114
Total fresh water inputs ($m^3 x 10^6$)	1 399	6 004	6 770
Evaporation ($m^3 x 10^6$)	2 565	1 890	1 529
Balance ($m^3 x 10^6$)	-1 166	+4 114	+5 241

In summary, the salinity of Lake Manzala was largely controlled by evaporation losses and compensating saltwater inflows during the 1920s and 1930s but by the 1960s drainwater inputs became the predominant factor and have become increasingly more so.

A comparison of nutrient and oxygen values during the 1960s with those of the 1979-80 indicates a pattern of increasing eutrophication and dominance of the large agriculture drains on the ecology of Lake Manzala.

Levels of nitrates and phosphates were relatively high and constant during 1962 to 1968 for all regions. Annual mean values of phosphate ranged from 0.021 to 0.105 mg/L, while nitrates varied from 0.159 to 0.634 mg/L. Phosphate levels in the southern sector were roughly two to three times those of other regions, while nitrate levels were only slightly higher. Presumably, this reflects the nutrient input to the southern sector via the drain at that time. During 1979, phosphate levels were approximately three times those of the 1960s, ranging from 0.050 mg/L in the western sector to 0.294 mg/L in the southern sector. The magnitude of change was highest in the southern sector and declined with the degree of influence of drainage water in the other regions. The pattern of nitrates is not clear, with the exception of a four-fold increase from the 1960s to 2.172 mg/L in the southern sector during 1979. The relative importance of the Bahr El Baqar Drain, which has flow rates of roughly half the Hadous Drain, is demonstrated by the fact that it has phosphate concentrations of six times and nitrate concentrations of about four times those of the Hadous and Ramsis Drains.

Mean oxygen concentrations during 1979-1980 were at supersaturation levels for most regions and are generally higher than those in the 1960s with the exception of the southern sector, where a depressed mean saturation value of 70 percent was recorded. Of signficance from a fishery viewpoint is the low level of oxygen recorded during July, 1979 (32 percent saturation or less than 3 mg/L). Such concentrations may be marginal for mullet and other less tolerant species. This is reflected by the virtual absence of mullet and other marine species from the catch in this region (0.08 percent in 1979 to 1980). However, the assimilation capacity of the system (i.e. the ability to absorb large amounts of nutrients and organic material) remains high. The Bahr El Baqar is completely anoxic (i.e. devoid of oxygen) until its mouth, but within 0.25 km, oxygen levels may be at saturation and nutrient levels reduced to a fraction of drain levels. This is the result of a very high rate of plankton production.

CATCH AND EFFORT

The mean number of licensed fishing boats roughly doubled between 1907 to 1935 and the early 1960s; from 1,300 to 1,400 boats to 2,500 to 2,600 boats. During the 1979-1980 study period the number of fishing boats was estimated at 4,010 (not including trading vessels). The number of licensed fishermen during 1928 to 1935 averaged 7,000 to 8,000, compared with 12,000 to 14,000 in the 1960s and 1970s. The estimated number of fishermen during 1979-1980 was 17,000 including 4,000 pre-adolescent boys (MacLaren 1981).

A summary of fishing gear and methods used on Lake Manzala during the past 50 years reveals that the two main methods have predominated throughout this period:

seine nets, which are especially effective for mullet species and trammel nets which are most effective for tilapia species. The predominance of seine nets methods during the 1920s reflects the importance of mullet species at that time, as does the high percentage of larger boats which were needed to handle these nets. In 1928, the number of trammel nets fished during the summer and winter was relatively constant (250 and 290 respectively), but the number of seine nets declined from 590 in summer and fall (when mullet were most abundant) to 262 in winter. This suggests that many of the seine net boats concentrated on mullet during the summer and did not bother to fish in winter. The number of seine nets and large boats recorded during 1979 compare with those of the 1920s. thus the increase in effort has been largely geared to expansion of the tilapia fishery (i.e. in terms of smaller boats and the use of trammel nets).

Catch statistics from 1907 to 1979 are plotted against time in Figure 2. This graph indicates catches during the period 1907 to 1960 ranged between 10,000 to 15,000 tonnes, although there is a 20 year gap in data between 1936 and 1956. Catches increased to about 20,000 tonnes during the mid-1960s and doubled to about 40,000 to 45,000 tonnes by the late 1970s. The open-water catch figure of 41,000 tonnes, based on catch survey results for 1979 to 1980, does not include the hosha production estimated at 15 20,00 to 25,000 tonnes. Hosha have become increasingly predominant since 1967 and it is unlikely that they made a significant contribution to the catch prior to this time.

The increasing importance of the Port of Matariya, located adjacent to the southern sector as a fishery centre is evident from records of fish landed at the main centres on Lake Manzala from the 1920s through the 1970s. The percentage of the total catch landed at Matariya increased from 42 percent during 1922 to 1933, to 85 percent to 1979 to 1980, or from 5,300 to 34,600 tonnes. This increase can be largely attributed to the expansion of the tilapia fishery in southern and eastern sectors of the lake, resulting from high nutrient inputs from the southern drains.

Historical seasonal catch trends reflect the change from a mullet fishery to one dominated by tilapia. During 1921 to 1933 about 70 percent of the fish was caught during August to January, when the young-of-the-year mullet became vulnerable to the fishery. Tilapia were also caught during this period as incidental species, but fishing activity was largely dictated by

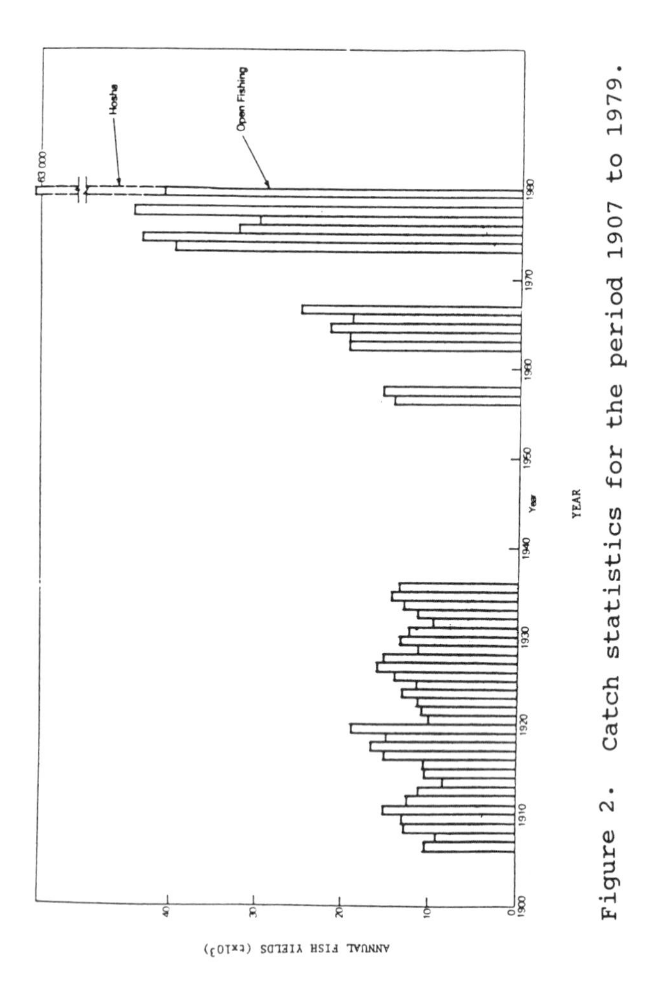

Figure 2. Catch statistics for the period 1907 to 1979.

the availability of mullet. The main change from the earlier period compared to the period 1957 to 1966 is a decrease in the November to January catch, and an increase in the May to July catch to levels equalling the winter months. This reflects the increasing tendency for fishermen to deliberately fish for tilapia during late spring and early summer. Apparently, the decline of high value mullet species in the catch had the effect of depressing fishing during the cold and windy period of November to January. The 1979 to 1980 data indicate a further shift of fishing activity from the winter months to spring and summer. Thus, fish catch is now more evenly distributed throughout the year than what it was in earlier historical periods. The periods of peak and low supply have been reversed.

Fish catches on Lake Manzala have been dominated by mullet and tilapia species over the past 70 years (Table 6). The dominant trend is a decline in marine species and an increase in tilapia. Mullet declined from 56 percent of the catch (7069 tonnes) in 1920 to 1929 to 22 percent (3,259 tonnes) in 1957 to 1958. There was a gradual decline through 1960s and early 1970s to 9.2 percent (2,219 tonnes) during 1972 to 1976 followed by an accelerated decline to 2.2 percent (902 tonnes) in 1979 to 1980. Of the four or five species of mullet recorded, Liza ramada has historically averaged 80 to 90 percent of the total mullet catch. Tilapia increased from 20 percent of the catch in 1920 to 1929 (2,521 tonnes) to 60 percent (8,894 tonnes) in 1957 and 1958 and to 64 percent (13,433 tonnes) in 1962 to 1966. There was a rapid increase during the 1970s (the figure of 19,771 tonnes for 1972 to 1976 is probably too low, see MacLaren 1981) to 85 percent (34,822 tonnes) in 1979 to 1980. Four species of tilapia are found in the commercial catch, but have not been distinguished in historical catch statistics. Wimpenny (1930) reported three species and stated that the salinity tolerant T. zillii was dominant. Bishara (1973) reported four species in 1967 of which S. nilotica was dominant. In 1979 to 1980 S. nilotica comprised 63 percent of the tilapia catch while T. zillii was less than 20 percent. The catch per fisherman remained relatively constant between the 1920s and the 1960s at 1.6 to 1.9 tonnes. The mean CPUE for Lake Manzala during 1979 was almost double the 1960 figure for fishermen at 2.5 tonnes. Considerable variation is found from region to region. The southern sector of the lake, which has the highest level of effort, or under one fisherman for every hectare, has the highest CPUE of 3.4. Catch per unit effort for the northern western and eastern sectors are roughly comparable to lake means for earlier periods, when

TABLE 6 SPECIES COMPOSITION OF CATCH IN RELATIVE PERCENTAGE AND TONNES FOR FIVE PERIODS FROM 1920 TO 1980 (MacLaren 1981, volume 10).

	Mullet	Crustaceans	Other Marine	Tilapia	Eels	Other Freshwater
1920 to 1929	55.8 (7 069)	7.7 (2 242)	3.3 (418)	19.9 (2 521)	3.1 (393)	0.06 (76)
1957 to 1958	21.9 (3 259)	5.9 (895)	7.2 (968)	59.7 (8 894)	1.1 (160)	4.4 (730)
1962 to 1966	13.4 (2 820)	9.7 (2 041)	6.4 (1 343)	64.0 (13 433)	0.7 (152)	5.7 (1 207)
1972 to 1976	9.2 (2 219)	1.9 (435)	1.4 (347)	82.0 (19 771)	0.9 (225)	4.6 (1 094)
1979 to 1980	2.2 (902)	6.0[a] (2 430)	0.7 (301)	85.4 (34 822)	0.8[b] (309)	4.9 (1 996)

[a]Of which 2 312 t (or 5.7 percent of total catch) is Palaeomon elegans.

[b]Not including an estimated 49 t of the inedible green eel Dalophis imberbis.

landings, and presumably catches were more evenly distributed around the lake. It is difficult to make regional comparisons of effort, as such data are not available for earlier periods. However, some idea of effort can be obtained by comparing total catch and CPUE.

Landing and catch data indicate that yields in the northern and western sectors have not changed much (100 to 150 kg/ha during 1979 to 80 versus mean catches of 100 to 170 kg/ha in the 1920s to 1960s). Constant catch and CPUE imply constant effort, while constant effort and CPUE are indicative of relatively stable stock conditions. Thus, it seems unlikely that there have been any significant stock increases in the western and northern sectors. Present catches in the eastern sector are high when compared to 1920 to 1960 values (350 to 500 kg/ha versus 100 to 170 kg/ha). Relatively constant CPUE and expanded catches imply increased effort and higher stock levels.

In summary, a comparison of total catch, effort and CPUE data seems to indicate the following:

-little change in catch and effort in northern and western parts of the lake during the past 70 years; moderate increase in catch and effort during the past 20 years in the eastern sector.

-large scale increases in catch and effort in the southern sector in recent years.

AN EVALUATION OF THREE FISHERIES PERIODS

The transformation from a marine-based fishery to one dominated by low value fresh water species has been apparent in Lake Manzala for some time. Frequent references are made to the negative impact of such change in the scientific literature and reports over the past two decades. Many fisheries officials and fishermen interviewed voiced similar sentiments. Thus fisheries management and development proposals and activities in recent years have been largely geared towards preserving and rehabilitating marine stocks (e.g. the clearing and construction of addition lake-sea connections to increase the salinity of lake waters and to facilitate the passage of marine species, especially mullet fry, into the lake). However, there is doubt whether such developments are necessary or beneficial in terms of overall fisheries performance.

The main components of the fish catch (historically tilapia and mullet) are likely to dominate in the future. The key water quality and environmental parameters (salinity, nutrient enrichment and fishing effort) are controllable to a degree, but optimal conditions for the two species groups appear largely incompatible under natural lake conditions.

An analysis of historical and current fisheries data provides us with a first level biological and socio-economic evaluation of three rather distinct open fishing periods (each associated with different periods in the lake's past development and each based on different patterns of salinity, nutrient inputs, fishing effort and species composition). One end of the spectrum is defined by a marine fishery dominated by mullet species. At the other end is fresh water fishery based on tilapia species. A third option is defined by intermedia conditions and a mixed stock. The key parameters of the three alternative scenarios are summarized in Table 7.

Period 1 is based on the years 1920 to 1929. Open lake area was 171,000 ha. Salinity levels were high averaging 14,000 mg/L. Drainwater and nutrient inputs

TABLE 7 KEY PARAMETERS OF THREE FISHERIES PERIODS (MacLaren 1981, volume 10).

	Period 1 1920 to 1929	Period 2 1962 to 1966	Period 3 1979 to 1980
Fish Yields (kg/ha)			
Mullet	50	23	14
Crustaceans (High Value)	15	16	2
Other Marine	3	13	4
Total High Value	68	52	20
Tilapia	18	106	535
Crustaceans (Low Value)	-	-	35
Other Fresh Water	3	8	36
Total Low Value	21	114	606
Total Yields (kg/ha)	88	167	626
Gross Value (LE/ha)	90	109	266
Physical and Chemical Characteristics			
Salinity (mg/L)	High (14 000)	Medium (8 000)	Low (3 000)
Nutrient Inputs	Low	Medium	High
Lake Area (ha)	171 000	126 000	89 000
Fishery Characteristics			
Total Number of Fishermen (per 100 ha)	7 700 (4.5)	13 000 (10.2)	17 000 (19.0)
Total Number of Boats (per 100 ha)	1 400 (0.7)	2 600 (2.1)	4 000 (4.5)

[a] *Palaeomon elegans*.

[b] Includes hosha portion of open lake production estimated at 15 000 t.

[c] Based on economic prices of LE400 and LE1200 per t for low value and high value species, respectively.

were low. Yields average around 88 kg/ha or 15,000 tonnes for the whole lake. Of the total catch, 77 percent (68 kg/ha) consisted of high value marine species. The bulk of this, or 50 kg/ha (56 percent of the total), was mullet species. Gross annual returns were estimated at LE 90 per hectare. Total fishing effort was relatively low (1,400 boats or 7,700 fishermen). However, CPUE was less than two tonnes per fisherman and on seven occasions during 1907 to 1935, annual yields declined from the previous year with an increase in effort,indicating possible overfishing at 12,000 to 18,000 tonnes. The mean yield figure of 15,000 tonnes is probably a reasonable estimate of sustainable yield during this period for the species assemblage that existed.

Period 2, representing the years 1962 to 1966 was one of environmental and fisheries transition. A reduction in lake area to about 126,000 ha and increased drainwater inputs reduced mean salinity levels to 8,000 mg/L. Total fishing effort almost doubled from the earlier period and yields increased to 167 kg/ha or 21,000 tonnes. Only 52 kg/ha (31 percent) were high value species of which 23 kg/ha (14 percent) consisted of mullet species. Gross returns from the open fishery were estimated at LE 69 per ha on an annual basis.

Period 3 represents the current (1979 to 1980) situation on Lake Manzala. Open lake area is now only 89,000 ha, mean salinity has been reduced to about 3,000 mg/L. Fish yields originating from the open water area are estimated at 626 kg/ha or 56,000 tonnes per annum. Of this total, 41,000 tonnes are caught by the open water fishery while an estimated 15,000 tonnes of the hosha production originates from the open water stock. The bulk of the catch (535 kg/ha or 85 percent) consists of tilapia species, while mullet species represent less than 15 kg/ha or about two percent. Annual gross returns per hectare of open lake are estimated at LE 266. Effort increased moderately to 4,000 boats or about 17,000 fishermen but the effort expressed in fishermen per hectare increased by 200 percent over that of Period 2, and 400 percent over that of Period 1.

The comparison of three fisheries periods indicates that an intensive, enriched tilapia-based fishery is vastly superior to either the mullet or the mixed species alternative according to the following criteria:

> Total yield: the total fish yield per unit area during Period 3 was about six times that of Period 1 and four times that of Period 2.
>
> Gross returns: gross returns per unit area from the fishery during Period 3 was two to three times that of Periods 1 and 2.
>
> Employment: employment in man years per hectare during Period 3 was about four and two times that of Periods 1 and 2, respectively.

Even with the reduction in lake size by one half from Period 1 to 3 the total yield increased by four while gross value of the lake fisheries doubled and fishing effort in terms of fishermen and boats more than doubled.

DISCUSSION AND CONCLUSIONS

Historically two groups of fish species, tilapia and mullet, have dominated the fishery on Lake Manzala. The transformation from a mullet to a tilapia based fishery has accompanied increased nutrient loading, a decline in salinity and increased exploitation pressure. The transformation reflects not only a relative increase in tilapia abundance but a real decline in mullet in terms of total catch. The historical analysis points out the basic incompatibility of a mullet fishery and an intensive tilapia system. This is confirmed by current yield, effort and water quality data. Mullet are virtually non-existent in the southern sector which is highly enriched by drainwater flows and where annual yield of tilapia species under intensive fishing effort may exceed 2,000 kg/ha (when both open water yields and hosha yields derived from open water stocks are considered). While most tilapia species can tolerate and survive over a wide range of salinities, on Lake Manzala, *Sarotherodon* species do not do well and are not abundant when salinities exceed 5,000 to 8,000 mg/L for extended periods and are virtually absent at salinities over 15,000 to 20,000 mg/L. Maximum tilapia productivity on Lake Manzala occurs at salinities below 3,000 mg/L.

The high yield of tilapia species, and in particular *S. nilotica*, reflects the ability of this group to benefit from high productivity associated with enrichment and its ability to cope with environmental instability and intensive exploitation pressure.

A high turnover tilapia fishery appears well suited to deal with pollution problems from pesticides and heavy metals associated with enrichment from agricultural drainage and sewage. Although levels of pollutants monitored were somewhat elevated in sediment and in some fish near the entrance of drains, recorded levels in water and fish were relatively low and all within accepted international health standards (MacLaren 1981, Vol. 7).

Fishing effort on Lake Manzala is not only intense but also very diverse and efficient in terms of habitats and fish species exploited. The requirement for regulatory management of the tilapia fishery would appear to be low.

FUTURE DEVELOPMENT

The intensive tilapia-based fishery has been identified as the optimal fisheries scenario in terms of yield, value and employment. Lake Manzala also serves an important role as nutrient assimilation and stabilization system for agricultural drainwater and for sewage outfall from Cairo. The shunting of the nutrient load directly into the colder Mediterranean would not only have a negative impact on the Lake Manzala fishery but would also be undesirable from an ecological and environmental point of view. The tourist beaches at Port Said located nearby would likely be affected during the summer because of prevailing northwest winds.

Nutrient loadings are expected to double by the year 2000. Based on existing nutrient loading, the extent of areas presently impacted by drainwater and elevated fish yields in these area, it is estimated that the fixation capacity of the Lake Manzala system will not be be exceeded and that there is potential for increasing fishery yields. The tilapia based model has been proposed as the basis for a comprehensive long term fisheries development plan for the lake. The plan calls for improved dispersion of drainwater and nutrients in Lake Um El Rish and in the southern, eastern and western sectors of Lake Manzala (MacLaren 1981, Vol. 6).

The Lake Manzala study indicated that nutrient inputs into the system were fixed rapidly by primary production processes. A high proportion of this production is converted to fish flesh by an assemblage of tilapia species which are then rapidly removed from the system by an intensive fishery. Any projection of incremental fish yield benefits from future increases in nutrient loading are highly speculative at this time. However, some first order relationships of nutrient input to fish yield can be developed. More than 80 percent of the annual total nutrient load (total nitrogen plus total phosphorus) estimated at 13,000 tonnes is loaded into the southern sector of Lake Manzala. Yields from the western and northern sector, where nutrient inputs are minimal, averages about 120 kg/ha. This is also a reasonable estimate of yield for earlier periods when the level of enrichment was low. If one assumes this to be a base yield level and that the balance in other lake areas is due to enrichment, total yields directly attributable to nutrient inflows are about 50,000 to 55,000 tonnes (including hosha areas and yields). This translates to a fish yield of roughly 4 - 5 tonnes per tonne of nutrient. On Lake Um El Rish, virtually the entire estimated annual yield of 3,500 tonnes originates from the northern end of the lake, which is exposed to Bahr El Baqar inflows. These balance the total evaporation from the lake estimated at 155 x 10^6 m^3. The total nutrient inputs from this inflow are estimated at 700 tonnes. This amounts to an incremental fish yield (assuming a base level of 120 kg/ha) of about 5 tonnes for each tonne of nutrients.

It is estimated that total nutrients will roughly double by the year 2000 to around 26,000 tonnes, a net increase of about 13,000 tonnes (MacLaren 1981, Vol. 6). Assuming a rate of four to five tonnes of fish per tonne of nutrients, fish yield could potentially increase from 50,000 to 65,000 tonnes on the main body of the lake providing that increased nutrient loads are effectively dispersed and that significant detrimental impacts on fish production from increased eutrophication do not occur.

ACKNOWLEDGEMENTS

The author would like to acknowledge the Government of Egypt, the United Nations Development Program and MacLaren Plansearch Inc. for their support and assistance.

REFERENCES

Arab Republic of Egypt and the Kingdom of the Netherlands. 1977. Lake Manzala, an ecological investigation into the environmental consequences of some Nile River and delta Development Projects. Haskonig Consulting Engineers and Architects, Nijnegen, Netherlands.

Bishara, N.F. 1973. Studies on the Biology of tilapia species in some lakes in U.A.R. Ph.D. Thesis, Cairo University, Cairo, Egypt.

Faouzi, G. 1936. Report on the Fisheries of Egypt for the year 1935. Ministry of Finance. Coastguards and Fisheries Service. Government Press, Cairo, Egypt.

Fouad, A.B. 1928. Report on the Fisheries of Egypt for the year 1927. Ministry of Finance. Coastguards and Fisheries Service. Government Press, Cairo, Egypt.

MacLaren Engineers Planners and Scientists Inc. 1981. Lake Manzala Study. Draft final report to the Arab Republic of Egypt and the United Nations Development Program. EGY/76/001-07, 12 volumes. Ministry of Development and New Communities, Cairo, Egypt.

Montasir, A.H. 1937. Ecology of Lake Manzala. Bulletin Faculty of Science 12. Cairo University, Cairo, Egypt.

Paget, G.W. 1924. Report on the Fisheries of Egypt for the year 1923. Ministry of Finance. Coastguards and Fisheries Service. Government Press, Cairo, Egypt.

Panse, V.G. and Sastry, K.V.R. 1960. Sample surveys for fishery statistics. Technical Report to the Government of the United Arab Republic. F.A.O. Report Number 1247, F.A.O. Rome, Italy.

Samra, A.F.A.G. 1933. Report on the fisheries of Egypt for the year 1931. Ministry of Finance. Coastguards and Fisheries Service. Government Press, Cairo, Egypt.

Shaheen, A.H. and Youssef, S.F. 1978. The effect of the cessation of Nile flood on the hydrographic features of Lake Manzala, Egypt. Arch. Hydrobiol. 85:(3):339-367.

Shaheen, A.H. and Youssef, S.F. 1979. The effect of the cessation of Nile flood on the hydrographic features of Lake Manzala, Egypt. Arch. Hydrobiol. 85 (2):166-191.

Toews, D.R. and Ishak, M.M. 1984. Fishery transformation on Lake Manzala, a brackish Egyptian delta lake, in response to anthropological and environmental factors during the period 1920-1980. General Fisheries Council for the Mediterranean. Studies and Reviews Number 61, Volume 1, F.A.O. Rome, Italy.

Wahby, S.D., Youssef, S.F. and Bishara, N.F. 1972. Further studies on the hydrography and chemistry of Lake Manzala. Bull. Inst. Ocean and Fish. 2:401-422.

Wakeel, S.K. and Wahby, S.D. 1970a. Hydrology and chemistry of Lake Manzala. Arch. Hydrobiol. 67(2): 173-200.

Wakeel, S.K. and Wahby, S.D. 1970b. Bottom sediments of Lake Manzala, Egypt. Jour. Sedimentary Petrology. 40(3):480-496.

Wimpenny, R.S. 1930. Report on the fisheries of Egypt for the year 1929. Ministry of Finance. Coastguards and fisheries Service. Government Press, Cairo, Egypt.

Youssef, S.F. 1973. Studies of the biology of family Mugilidae in Lake Manzala. M. Sc. Thesis, Cairo University, Cairo, Egypt.

CHAPTER 4

CONTAMINANTS IN SELECTED FISHES FROM THE UPPER GREAT LAKES

Frank M. D'Itri

Institute of Water Research and Department of Fisheries and Wildlife, Michigan State University, East Lansing, Michigan 48824-1222

ABSTRACT

As early as 1965 it was apparent that Great Lakes fishes were bioaccumulating relatively large concentrations of agricultural insecticides such as DDT and dieldrin while the bioaccumulation of PCBs was suspected, but not confirmed. By 1969 state and government agencies had begun annual organochlorine contaminant monitoring programs. They constitute the majority of data on Great Lakes fishes. In general, these data show that, between 1965 and 1980, the PCB, total DDT, dieldrin and mercury concentrations in lake trout, bloater chub, coho salmon and chinook salmon as well as other species of fishes have decreased significantly throughout the Great Lakes. This reflects the more stringent controls on point discharges of these contaminants into the water system in the 1970s. Since 1980, however, the concentrations of these contaminants in Great Lakes fish have declined only slightly in many cases, remained relatively constant, or even increased slightly in others. The recent data often exhibit such year-to-year variations that it is not possible to discern trends. One reason may be the uncontrollable variables inherent in a large lakes fish sampling program. Another may be the impact of toxic chemicals from diffuse non-point sources such as atmospheric deposition, municipal/industrial effluent discharges, agricultural and urban runoff, reactivation of contaminants from the sediments, and leachate from municipal and industrial landfills.

INTRODUCTION

Since Rachel Carson published 'Silent Spring' in 1962, awareness and concern have increased with the respect to the effects of a wide range of persistent chemicals on the environment and human health. At about the same time, the development of new and improved analytical detection methods led to the identification and quantification of organic chemicals at concentrations thousands of times lower than previously was possible. As a result, scientists were able to document not only the presence but also the cycling of many xenobiotic chemicals throughout the aquatic ecosystem.

For the past two decades, organochlorines have been identified as serious contaminants resulting in long-range detrimental effects on sports and commercial fishing in the Great Lakes that constitute part of the boundary between the United States and Canada. As early as 1965 it was apparent that Great Lakes fishes were bioaccumulating relatively large concentrations of agricultural insecticides such as DDT and dieldrin while the bioaccumulation of PCBs was suspected, but not confirmed. Consequently, discharges of major persistent chemicals were curtailed in the Lake Michigan watershed after 1968 when the Lake Michigan Enforcement Conference recommended restricting the use of DDT in Illinois, Wisconsin and Michigan.

Between 1969 and 1972 legislation was enacted to restrict or ban the use of dieldrin, DDT, PCBs and mercury as well as other toxic chemicals within the Great Lakes basin. Then state and federal governments on both sides of the border were generally quite effective in eliminating point discharges of toxic chemicals and controlling the uses of agricultural chemicals such as pesticides and herbicides. As a result, the PCB, total DDT, dieldrin and mercury concentrations in lake trout (*Salvelinus namaycush*), bloater chubs (*Coregonus hoyi*), coho salmon (*Oncorhynchus kisutch*), and chinook salmon (*Oncorhynchus tschawytscha*) as well as other species of fish appeared to decrease significantly between 1975 and 1980. Contaminant burdens in Lake Superior fish, for example, especially the top predator species, such as lake trout, brown trout (*Salmo trutta*), bloater chubs and whitefish (Coregonus culpeaformis) declined from the early 1970s on.

Another significant factor was the 1972 Agreement on Great Lakes Water between the United States and Canada. This modified the 1909 Boundary Waters Treaty and initiated a more intensive effort on the part of the two governments to reduce nutrient loadings and curtail inputs of certain toxic chemicals, particularly after the enactment of a second Great Lakes Water Quality Agreement (GLWQA) in 1978. It emphasized that the primary concern had shifted from excessive nutrient loading toward the control of toxic substances as part of an integrated ecosystem. While the goals of the 1978 GLWQA are still far from being completely attained, progress continues; and the concentrations of many persistent chemicals are lower now than they were in the early 1970s. However, one of the notable discoveries is that the levels of contaminants stabilized and sometimes even increased after the initial drop despite the elimination of point discharges.

The effectiveness of the initial lessening of point source discharges was demonstrated as various species were intensively monitored by state and federal agencies. Lake trout, coho salmon and walleye (Stizostedion vitreum) were monitored as top predator fishes likely to accumulate higher levels of contaminants, just as the herring gull, further up the trophic level, concentrated more than smaller fish and birds. Contaminant levels tended to be higher also in lake trout because this species is long-lived and because of the way lipophilic contaminants accumulate in its tissues (Eschmeyer and Phillips 1965).

The first federal effort to study the problem began in 1967 when the Bureau of Sport Fisheries of the United States Fish and Wildlife Service (USFWS) initiated the National Pesticide Monitoring Program (NPMP) to assess the general levels of pesticide residues in the environment. The results of whole-fish analyses confirmed that organochlorine insecticides like DDT and dieldrin were being accumulated in fish tissues collected nationwide. Similarly, in 1969-1970, the USFWS initiated an open lakes program not only to identify newly emerging contaminant problems but also to establish their long-term temporal and spatial trends in the Great Lakes. Since fish serve as biological integrators of xenobiotic contaminants, the researchers concluded that only whole-fish samples could accurately reflect their exposure to the chemicals. Therefore, a yearly whole-fish sampling program was established to be consistent with respect to location and method of collection of fish of the same species, size, age and sex.

At about the same time, the United States Food and Drug Administration (FDA) and public health departments in the states bordering the Great Lakes began programs to determine the spatial distribution of contaminants and to enforce the FDA tolerances for the consumption of contaminated food. These agencies undertook only required compliance monitoring to address public health concerns, not long-term research. They are statutorily required to perform only analyses to certify that the fish contain less than the legal tolerance level of a contaminant. Consequently, their analytical methods often are not sensitive enough to detect contaminants at lower but still ecologically significant levels. Moreover, because they were primarily interested in monitoring the contaminant levels in the edible portions of fish, the samples were prepared in the same manner that the fish are delivered to market, i.e. headless-gutted, skin-on fillet or skin-off fillet. The choice of the latter method is significant because the concentrations of some organochlorine compounds can be substantially higher (usually by a factor of 2 to 4) in a skin-on fillet sample compared with those with the skin removed. However, some health officials favored analyzing only skinless fillets not only because this generally produced lower contamination levels but also because it was much easier and, therefore, cheaper.

This policy decision is the primary reason that the enormous set of historical compliance monitoring data generated by the FDA and/or the state agencies are not directly comparable with the whole fish data produced by the USFWS and Canada Department of Fisheries and Oceans(CDFO). The data in this paper reflect reports/publications from the Great Lakes Environmental Contaminants Survey (GLECS): 1972-1980 (Forney 1982); the Ohio Department of Natural Resources (ODNR 1980); the Wisconsin Department of Natural Resources (WDNR 1985, Sheffy 1977, Sheffy and St. Amont 1980); the Indiana Department of Natural Resources (Lauer 1980); the Michigan Departments of Natural Resources (Rybicki and Keller 1978, Powers 1976, Hartig and Sifler 1980, MDNR 1979), Agriculture (MDA 1975a, 1975b, 1976), and Public Health, the Michigan Water Resources Commission (MWRC 1975); the U.S. Fish and Wildlife Service (Henderson et al. 1969, 1971, 1972; Inglis et al. 1971; May and McKinney 1981; O'Shea and Ludke 1979; Schmitt et al. 1981, 1983, 1985; Walsh et al. 1977), the U.S. Environmental Protection Agency (Rockwell et al. 1980, GLNPO 1981); and the U.S. Food and Drug Administration (Hoeting 1983) and the International Joint Commission (IJC 1977, 1978a, 1978b, 1981, 1983).

The Canadian fish contaminant data were obtained from reports and publications from Environment Canada (1972-1977), Fisheries and Oceans Canada (1977-1980) and the Ontario Ministry of the Environment (OMOE 1976, 1977). Wherever possible, fish contaminant data were also summarized from the literature. (Frank et al. 1978, 1979, 1981, Rohrer et al. 1981, 1982, Clark et al. 1984, Whittle 1985, DeVault 1985, DeVault et al. 1985, Willford et al. 1976, DeVault and Weishaar 1983, 1984, Rodgers and Swain 1983, Armstrong and Lutz 1977, Jaffe et al. 1984, Jensen et al. 1982, Kreis and Rice 1985, Kuehl et al. 1983, Lowe et al. 1985, Murphy and Rzeszutko 1977, Neidermyer and Hickey 1976, Olsson et al. 1979, Parejko 1975, Rossman 1976, Reinert 1970, Reinke et al. 1972, Safe 1984, Simmons 1984, Strachan and Gloss 1978, Swain 1978, Veith and Lee 1971, Veith 1975, Veith et al. 1977, 1979, 1981, Wszolek et al. 1979, Young et al. 1972).

While a wide variety of fish were analyzed over the twenty years from 1966 through 1986, in each of the Great Lakes two or three species were analyzed with sufficient frequency to demonstrate some patterns if not trends. Usually these fish were selected because they were economically important to the fishery or were expected to be the most contaminated and so were favored as indicator organisms. In Lake Superior the lake trout and siscowet subspecies were most commonly analyzed; in Lake Michigan they were lake trout and bloater chub. These two were also commonly analyzed in Lake Huron along with whitefish. Numerous samples of yellow perch (Perca falvescens) and walleye as well as muskellunge (Esox masquinongy) were analyzed from Lake St. Clair. While Coho salmon are most important in Lake Michigan, enough analytical results were available to compare them for Lake Huron as well. During 1979, coho and chinook salmon from Lake Michigan and the Platte and Little Manistee Rivers were analyzed for DDT, dieldrin and PCBs. The results showed all salmon to be below the FDA action levels for each of the contaminants (Hartig and Stifler 1980). In 1980, the survey area was expanded to include the Grand and St. Joseph Rivers (Rohrer et al. 1982).

While the total DDT and dieldrin concentrations in both coho and chinook were below the FDA action level, the PCB levels, which ranged from 3.0 to 6.4 mg/kg in the chinook and 1.88 to 4.4 mg/kg in the coho, were in excess of the current FDA action level of 2.0 mg/kg. Coho salmon surveyed at 12 sites throughout the Great Lakes basin showed PCB, total DDT and dieldrin levels of 1.90, 0.67 and 0.06 mg/kg in 1980 and 0.79, 0.36 and 0.02 in 1982 (Clark et al. 1984, DeVault and Weishaar 1984).

Whether the fish samples were taken in open waters of the Great Lakes or in one of the basins, harbors or tributary outlets also significantly affected the levels of pollutants in fish because many of these areas are exceptionally highly contaminated with organochlorine compounds, especially PCBs. Ninety percent or more of resident fish (primarily carp, Cyprinus carpio) collected between 1981 and 1984 from 30 major harbors and tributaries contained PCB and dieldrin levels which exceeded the FDA action level. In addition, extremely high levels of PCB were observed in fish from the Kinniekinnie, Milwaukee, Fox (10-30 mg/kg) and Lower Sheboygan (> 90 mg/kg) Rivers in Wisconsin, while in Michigan the levels of PCB in fish from White Lake near Muskegon and the St. Joseph and Kalamazoo Rivers ranged between 10 and 30 mg/kg (Table 1).

For this report the fish contaminant data sets from a wide variety of sources were plotted as a function of time to produce series graphs that demonstrate the long-term trends of PCBs, dieldrin, and DDT in the indicated fish from the Great Lakes. This method combines data over a range of fish sizes, ages and lipid contents as well as variable sample types such as whole fish, skinless, or skin-on fillets and with different analytical protocols. While the disadvantage is that this method lacks confidence limits, this averaging of data reflects the fact that, in a given population of fish, the contaminant may vary by a factor of 5 or 6 as a function of the fish's migratory feeding, lipid, metabolic, or genetic and/or abiotic exposure differences. Other factors contributing to fish data variability are presented in Table 2.

LAKE SUPERIOR

Some of the earliest contaminant burden data for Lake Superior fish were collected between 1965 and 1968 (Reinert 1970). The mean total DDT levels were: 1.09 mg/kg in bloater, 1.02 mg/kg in coho, 0.45 mg/kg in whitefish and 0.21 to 7.44 mg/kg in several sizes of lake trout. The corresponding dieldrin levels rangedbetween 0.01 and 0.05 mg/kg. After the United States Fish and Wildlife Service National Pesticide Monitoring Program (NPMP) was initiated in 1967 to assess the general levels of pesticides in the environment nationwide, they reported the total DDT in lake trout to range between 0.80 and 1.38 mg/kg. The corresponding dieldrin levels were also low, ranging from 0.01 to 0.06 mg/kg (Henderson et al. 1969). A

TABLE 1 TOTAL PCB, TOTAL DDT AND DIELDRIN CONCENTRATIONS (mg/kg wet weight) IN FISH FROM SELECTED GREAT LAKES HARBORS AND TRIBUTARIES[1]

SPECIES	RIVER	TOTAL PCB mg/kg	TOTAL DDT mg/kg	DIELDRIN mg/kg	YEAR
		OHIO			
Carp	Black	1.28	0.26	0.02	1981
Northern Pike	Ashtabula	6.58	0.60	0.01	1981
Blue Gill	Ashtabula	10.68	0.09	0.01	1981
Brown Bullhead	Ashtabula	2.98	0.03	0.06	1981
Yellow Bullhead	Ashtabula	1.72	0.02	<0.002	1981
		WISCONSIN			
Carp	Sheboygan	38.60	0.88	<0.002	1980
Carp	Sheboygan	98.44	1.48	<0.002	1980
Northern Pike	Sheboygan	63.14	1.93	<0.002	1980
Black Crappie	Milwaukee	15.54	0.31	0.02	1981
Red Horse Sucker	Milwaukee	6.63	0.15	0.005	1981
Carp	Menominee	3.20	1.70	0.02	1981
Bullhead	Menominee	0.75	0.26	0.01	1981
Carp	Kinniekinnie	17.73	1.58	0.15	1981
Carp	Wolf	0.18	0.22	0.01	1980
Walleye	Wolf	0.84	0.14	0.01	1980
Carp	Fox Ab. DePere	9.53	0.50	0.09	1980
Carp	Fox Ab. DePere	7.22	0.11	0.02	1981
Rock Bass	Fox Ab. DePere	2.01	0.08	<0.002	1980
Carp	Fox Ab. DePere	20.89	0.27	0.02	1980
Walleye	Fox Ab. DePere	8.61	0.65	0.03	1980
White Sucker	Chequamegon Bay	0.66	0.26	0.01	1980
Walleye	Chequamegon Bay	0.40	0.20	<0.002	1980

[1] Data summarized from DeVault 1985

second NPMP survey in 1969 generated similar but slightly higher values (Henderson et al. 1971).

A third NPMP survey, conducted between 1970 and 1974, (Henderson et al. 1972), reported generally increasing total DDT levels for lake trout ranging from 0.80 mg/kg in 1970 to 5.14 mg/kg in 1974 and, for the first time, PCBs which also showed an increasing trend from 1.29 mg/kg in 1970 to 4.50 mg/kg in 1974 (Schmitt et al. 1983). The mean total whole lake trout levels of DDT were 3.54 mg/kg with a maximum value of 11.54 mg/kg (Poff and Degurse 1970). Concentrations of 1.32 mg/kg for DDT residues and 0.06 mg/kg for dieldrin were

TABLE 2 FACTORS CONTRIBUTING TO FISH DATA VARIABILITY

FISH
- Species
- Size (length/weight)
- Age
- Lipid Content
  (time/inter and intra species differences)
- Feeding Characteristics
- Migration/Mobility Characteristics
- Sex
- Respiration Rate
  (species and temperature differences)

SAMPLE COLLECTION
- Number of Fish (variable composites)
- Method (gill net/trap net/shocking)
- Sampling Location
- Sampling Time (season)

SAMPLE TYPE
- Fillet (skin-on/skin-off)
- Whole Fish

SAMPLE PREPARATION
- Storage (freezer dehydration)
- Time (lysosomal/bacterial decomposition)
- Packaging Material (plastics)

ANALYTICAL METHODS
- Analysis Protocol
- Interferences
- Quantification Procedure
- Quality Control
- Data Handling/Processing

UNCONTROLLABLE FACTORS
- Weather Events (spring runoff)
- Abiotic Environment
  (dissolved and undissolved humics/turbidity)
- Budgetary Constraints
  (results in data gaps which impact trend analyses)
- Degree of Industrialization (changes with time)
- Water Quality of Receiving Stream
  (less/new contaminants over time)

reported in an unspecified number of lake trout fillets collected in the vicinity of Thunder Bay and Marathon, Ontario (Reinke et al. 1972). Lake trout fillets collected around Isle Royale and Caribou Island in 1971 (Parejko et al. 1975) showed average total DDT residue levels of 3.25 mg/kg whereas lake trout collected near the Apostle Islands in 1972 and 1973 contained 3.7

mg/kg DDT residues, 1.8 mg/kg PCBs and 0.3 mg/kg dieldrin (Veith et al. 1977). In 1972-1973, the Fish Inspection Branch of the Canadian Department of Fisheries and Environment analyzed fillets of lake trout collected along the Canadian shore. They reported mean and maximum DDT residues of 1.28 and 7.35 mg/kg; PCB of 2.23 and 9.50 mg/kg; and dieldrin of 0.08 and 0.48 mg/kg (Williams 1975). In 1973, the IJC upper Lakes Reference Group detected the presence of other persistent compounds such as PCB, hexachlorobenzene, heptachlor, methoxychlor and toxaphene. In 1974, the PCB and DDT levels in Lake Superior lake trout were reported at 4.38 and 2.02 mg/kg, somewhat lower than in the 1974 NPMP survey (Armstrong and Lutz 1977).

In the spring of 1972, the Great Lakes Environmental Contaminants Survey (GLECS) was initiated to monitor and assess the concentrations of toxic contaminants in Great Lakes fish (Forney 1972). Although chub and whitefish were included, the emphasis was on lean lake trout, and Siscowet lake trout which, because of their greater lipid content, concentrate higher levels of PCBs and other lipophilic contaminants. The survey demonstrated that dieldrin has had minimal impact upon fish in Lake Superior. Of all lake trout sampled under the GLECS program, only one collected in 1976 exceeded the FDA action level of 0.3 mg/kg for dieldrin. However, in 1972, data indicated that the PCB and DDT levels were consistently above FDA action levels for these compounds. By 1980, analyses showed that DDT had rapidly declined throughout the lake in both subspecies of lake trout, and none exceeded the then FDA action level of 5.0 mg/kg. In 1980, neither subspecies of trout registered higher than 1.0 mg/kg. However, PCB contamination still presents a significant problem. While the levels have decreased since the beginning of the GLECS program, 8 percent of the common lake trout and 21 percent of the siscowet lake trout taken in 1980 exceeded the then 5.0 mg/kg FDA action level.

The data from these and other early studies of the contaminant concentrations in Lake Superior fishes are important background to evaluate more recently acquired information. The historical data suggest that the PCB, total DDT, and dieldrin levels declined after 1974-1975; and this result may be directly correlated with the banning or restricted uses of these compounds in the early 1970s (Figures 1, 2 and 3).

The U.S. Environmental Protection Agency Great Lakes National Program Office (GLNPO) and the U.S. Fish and Wildlife (FWS) Great Lakes Fishery Laboratory (GLFL) began trend monitoring in Lake Superior in 1977 followed by the Great Lakes Fisheries Research Branch

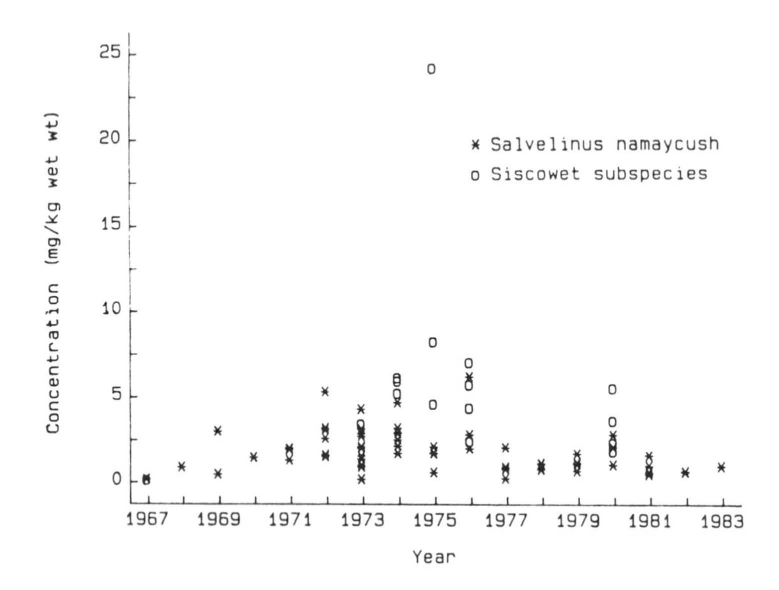

Figure 1. PCB Residues (mg/kg wet weight) in Lake Superior Lake Trout: 1967-1983.

(GLFRB) of the Canada Department of Fisheries and Oceans (CDFO) in 1980. However, the large data variability and brief study period, 5 and 3 years, have limited the detection of meaningful trends. Additional years of collection and analysis will be required for this. Moreover, the relatively low levels of PCB and dieldrin in the GLNPO-FWS data did not lend to the exhibition of trends or statistically significant ($P < 0.05$) differences over the study period (DeVault et al. 1985). The total PCB concentrations reported as Aroclor 1254 have been less than 2 mg/kg since monitoring began in 1977. Despite apparent year-to-year variability, the 0.5 mg/kg registered in 1982 is the lowest so far. The dieldrin concentrations have remained consistent and low, ranging between 0.05 and 0.005 mg/kg. The study by the Canada Department of Fisheries and Oceans (Whittle 1985) did, however, show that in aged 4+ lake trout the PCB levels declined steadily from 0.90 mg/kg in 1980 to a significantly lower level, 0.38 mg/kg in 1982 and then increased to 0.52 mg/kg in 1983.

The DDT concentrations in Lake Superior lake trout showed a significant decline. The levels reported by GLNPO-FWS declined from 1.2 to approximately 0.5 mg/kg between 1977 and 1982. In the Canadian study, between

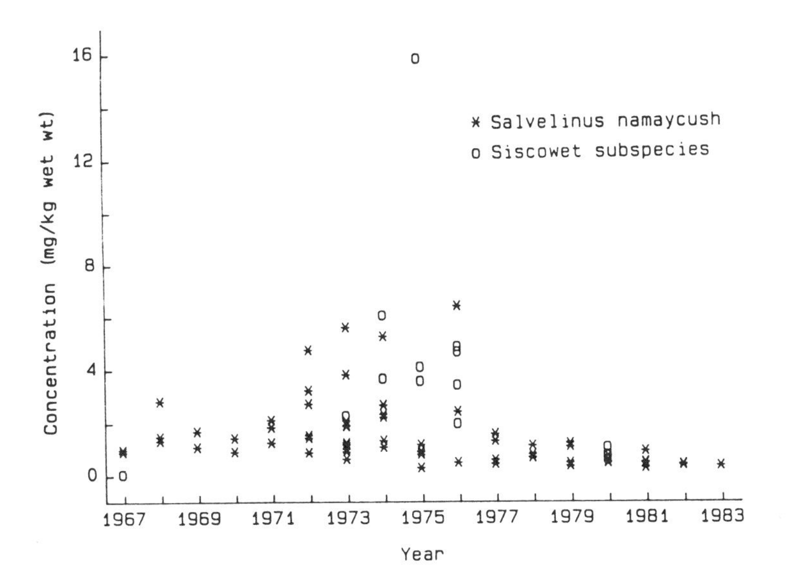

Figure 2. Total DDT Residues (mg/kg wet weight) in Lake Superior Lake Trout: 1967-1983.

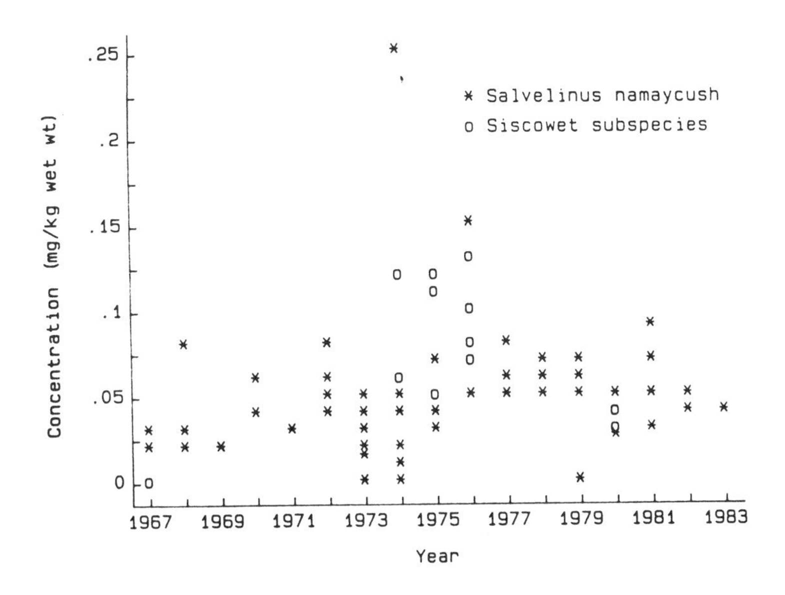

Figure 3. Dieldrin Residues (mg/kg wet weight) in Lake Superior Lake Trout: 1967-1983.

1980 and 1983, the total DDT levels decreased from 0.35 to 0.15 mg/kg, while the dieldrin concentration fluctuated between 0.03 and 0.07 mg/kg over the same period. In general, the contaminant levels in Lake Superior fishes were consistently the lowest relative to the same species in the other lakes with the PCB levels less than 50 percent of those in fish from any of the other Great Lakes.

Siscowet lake trout usually have higher mercury levels than the common lake trout collected from the same site in the same year. While the mercury levels in the lake trout and siscowet subspecies have declined below the current FDA action level of 1.0 mg/kg, the decline in

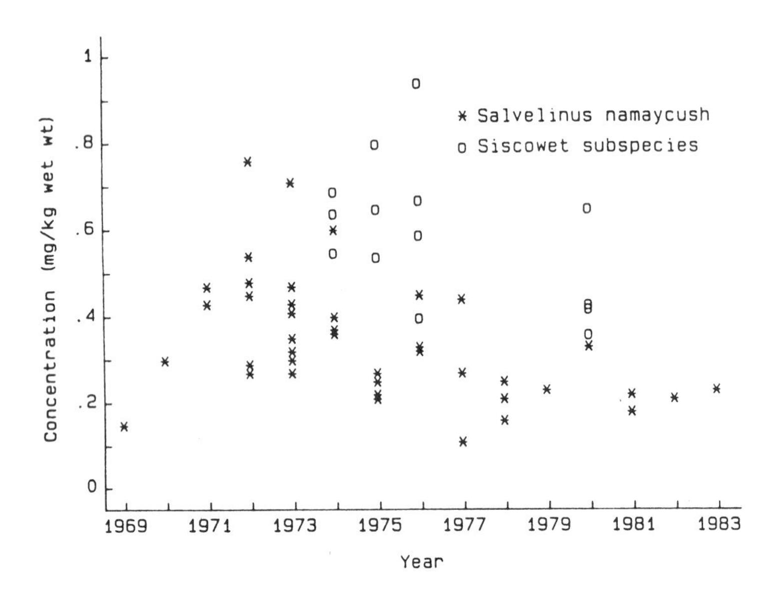

Figure 4. Mercury Residues (mg/kg wet weight) in Lake Superior Lake Trout: 1969-1983.

the siscowet was not as great. They still frequently exceeded the Michigan Department of Public Health advisory level of 0.5 mg/kg. The 1980-1983 mercury levels in the common and Siscowet lake trout ranged between 0.2 and 0.5 mg/kg in recent years. These values are consistent with trends observed in the other Great Lakes (Figure 4).

LAKE MICHIGAN

In the fall of 1971, thirteen species of fish taken from fourteen areas of Lake Michigan were analyzed for PCBs and DDT analogs (Veith 1975). The mean wet weight levels of PCBs similar to Aroclor 1254 ranged from 2.7 mg/kg in rainbow trout (Salmo gairdneri) to 15 mg/kg in lake trout. Most trout and salmon longer than 30 cm contained more than 5 mg/kg, the FDA action level guideline prior to 1980. The concentrations of PCBs in Lake Michigan coho salmon were 2 to 3 times greater than in Lake Huron, approximately 1.5 times greater than in Lake Ontario, and approximately 10 times greater than in coho from Lakes Erie and Superior. In addition, essentially 100 percent of the large salmon and trout and between 50 and 80 percent of the bloater chubs contained PCB levels greater than 5 mg/kg. Mean concentrations of total DDT ranged from less than 1 mg/kg in suckers (Catostomus spp) to approximately 16 mg/kg in large lake trout.

Since then, the trends for all contaminants in Lake Michigan have been generally downward. Under the 1985 advisory, coho, steelhead, and small lake trout joined yellow perch and smelt (Osmerus mordax) in the lowest category of concern. While analyses of bloater chubs and lake trout collected off Saugatuck, Michigan, indicate substantial declines in total DDT and PCB, these compounds still frequently exceed the current FDA action level of 2.0 mg/kg for PCB, particularly in large lake trout and bloater chubs. In addition, the levels of DDT, dieldrin and PCBs are consistently higher in fish taken from the southern end of the lake. These correspond closely with the higher levels of contaminants in the sediments.

Historically, lake trout, salmon and bloater chubs have had higher levels of PCB than whitefish taken in the same year. The PCB levels in bloater chubs have steadily decreased from about 5.6 mg/kg in 1972 to 2.1 mg/kg in 1982, still slightly more than the FDA action guideline of 2.0 mg/kg (Figure 5). The 1983-1984 Michigan GLECS data average 1.1 mg/kg with 2 of the 12 fish containing more than 2.0 mg/kg (Forney 1982).

The PCB concentrations in lake trout increased from 12.86 mg/kg in 1972, the first year that whole fish contaminant body burdens were measured, to a maximum of 22.91 mg/kg in 1974, followed by a gradual decrease to 8.18 mg/kg in 1978. Once again, the PCB concentrations apparently increased to 8.82 mg/kg in 1979 and peaked at 9.93 mg/kg in 1980. Since then, the levels have slowly decreased to 5.63 mg/kg in 1982. The 1984 data show that lake trout shorter than 51 cm decreased to the point where 90 percent recorded less than the FDA action level of 2.0 mg/kg. However, PCB concentrations in large lake trout as well as coho and brown trout longer than 64 cm remain high and should not be consumed (Figure 5).

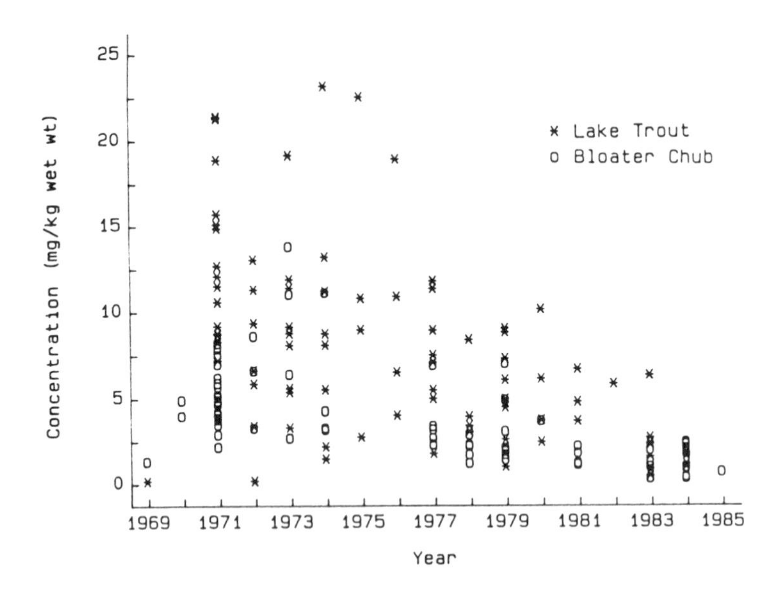

Figure 5. PCB Residues (mg/kg wet weight) in Lake Michigan Lake Trout and Bloater Chub: 1969-1985.

The USEPA Great Lakes National Program Office and the USFWS Great Lakes Fishery Laboratory compiled the only data set that included continuous contaminant body burdens in lake trout from Lake Michigan since 1970. These data show that the mean PCB concentrations initially increased from 12.86 mg/kg in 1972 to 22.91 mg/kg in 1974 and then declined to 5.63 mg/kg in 1982, an overall decrease of 75 percent (Figure 6).

Of the data sets for the three upper lakes, only those for Lake Michigan were sufficient to allow rigorous statistical evaluation of trends over time. Because the data sets for Lake Michigan were begun in 1972, DeVault et al. (1985) were able to show that the mean PCB and mean total DDT concentrations in lake trout taken off Saugatuck, Michigan, declined in a manner that approximated first order loss kinetics. These findings are similar to the decline of PCB

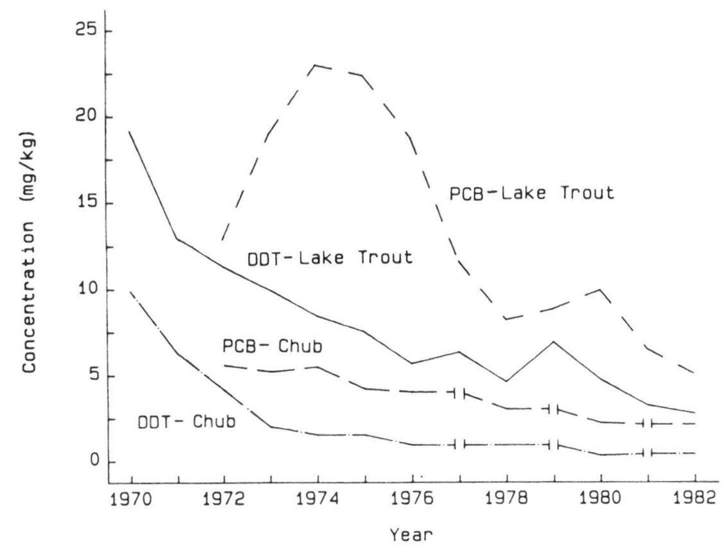

Figure 6. PCB and Total DDT Residue (mg/kg wet weight) Trend Analyses of Whole Body Lake Michigan Lake Trout and Bloater Chub: 1970-1982. Data from DeVault et al. 1985.

concentrations in bloater chubs taken near Saugatuck between 1972 and 1980 (Rogers and Swain 1983). They also approximated first order loss kinetics. Rodgers and Swain (1983) forecast that if the PCB loading into Lake Michigan continued to decrease at a rate at least equal to that before 1982 and if these declines continued at the current rate, the PCB levels in lake trout would reach 2.0 mg/kg in 1987. DeVault et al. (1985) projected that this level would be reached in 1988. This may be difficult to attain because atmospheric deposition now is estimated to contribute over 50 percent of the total PCB loading to the lake (Murphy and Rzeszutko 1977).

The mean total DDT concentrations in lake trout decreased from 19.19 mg/kg in 1970 to 2.74 in 1982, an overall 85 percent decrease that also followed first order loss rate kinetics (Figure 7). DeVault et al. (1985) predicted that if these declines continue at the present rate, the total DDT concentrations in lake trout will reach the IJC Agreement objective of 1.0 mg/kg by 1991. On the other hand, dieldrin levels in the larger lake trout and bloater chubs frequently exceed the current FDA action level of 0.3 mg/kg; and these levels have shown a tendency to increase since 1970 (Figure 8). The mean dieldrin concentrations in lake trout decreased from 0.27 mg/kg in 1970 to 0.20 mg/kg in 1971 and then increased over eight years to a maximum of 0.58 mg/kg in 1979. Then the concentrations declined to 0.21 mg/kg in 1982. This decrease also followed first order loss kinetics. The mean dieldrin concentrations in the 1983-1984 GLECS sampling were about 0.19 mg/kg (Forney 1982). However, dieldrin residues resulting from aldrin applied to crops before it was banned in 1974 are still being transported from fields into the aquatic ecosystem (Leung et al. 1981, Ricci et al. 1983, Schnoor 1981).

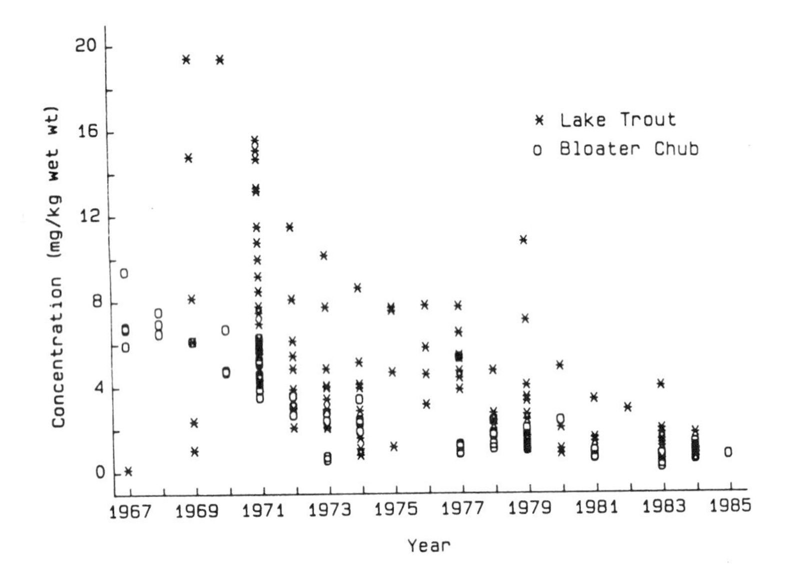

Figure 7. Total DDT Residues (mg/kg wet weight) in Lake Michigan Lake Trout and Bloater Chub: 1967-1986.

After the dieldrin levels appeared to decrease from about 0.26 to about 0.18 mg/kg in bloater chubs in 1972, they increased to a maximum of 0.55 mg/kg in 1978. Since then, the levels appear to have decreased to about 0.42 mg/kg in 1982, still in excess of the USFDA action guideline of 0.3 mg/kg (Figure 8). However, the mean dieldrin concentration in the 1983-1984 GLECS sampling of 12 bloater chubs was 0.26 mg/kg, slightly less than the FDA action guideline (Forney 1982).

The reason for the increase in the mean concentrations of dieldrin in lake trout between 1979 and 1982 is not

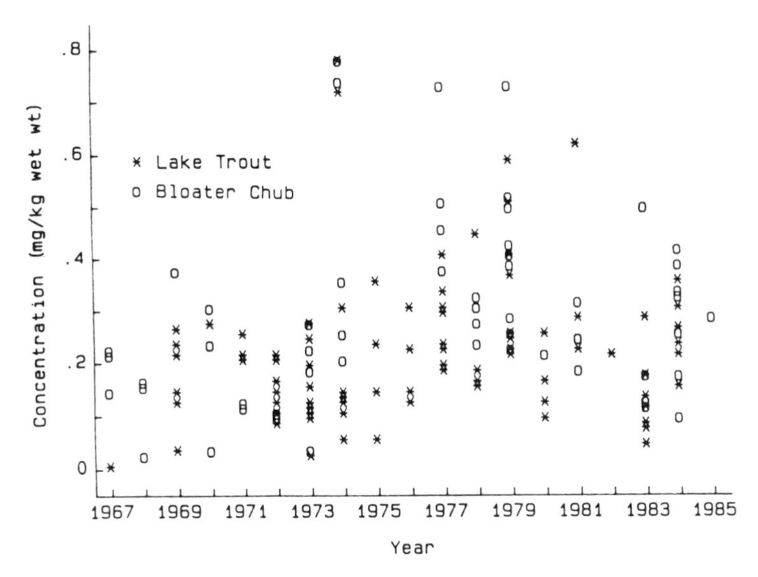

Figure 8. Dieldrin Residues (mg/kg wet weight) in Lake Michigan Lake Trout and Bloater Chub: 1967-1985.

clear as most of the uses of this compound were halted in 1974. Moreover, while 'the mean dieldrin levels in lake trout decreased between 1980 and 1982, the mean concentrations in other species such as bloater chubs and rainbow smelt did not. Dieldrin levels in bloaters increased from a mean of 0.27 in 1970 to 0.43 mg/kg in 1982 (Willford 1982) while the rainbow smelt remained relatively constant, ranging between 0.04 and 0.071 mg/kg from 1977 to 1982. Consequently, additional years of monitoring will be required to establish if the decline in dieldrin concentrations between 1979 and 1982 truly represents a trend.

Mercury levels in fish from Lake Michigan have been declining since 1972. Since 1980 they have been consistently below the FDA action level of 1.0 mg/kg although about 6 percent of the lake trout and chinook salmon exceeded the Michigan Department of Public Health's 0.5 mg/kg advisory level (Figure 9). The lakewide mean mercury levels by species for 1980 were: 0.22 mg/kg for lake trout; 0.35 mg/kg for chinook salmon; 0.17 mg/kg for coho salmon; 0.04 mg/kg for whitefish; and 0.10 mg/kg for bloater chubs (Rohrer 1982). Clark et al. (1984) reported 1980 mercury levels in five Lake Michigan tributaries ranging between 0.07 and 0.20 mg/kg.

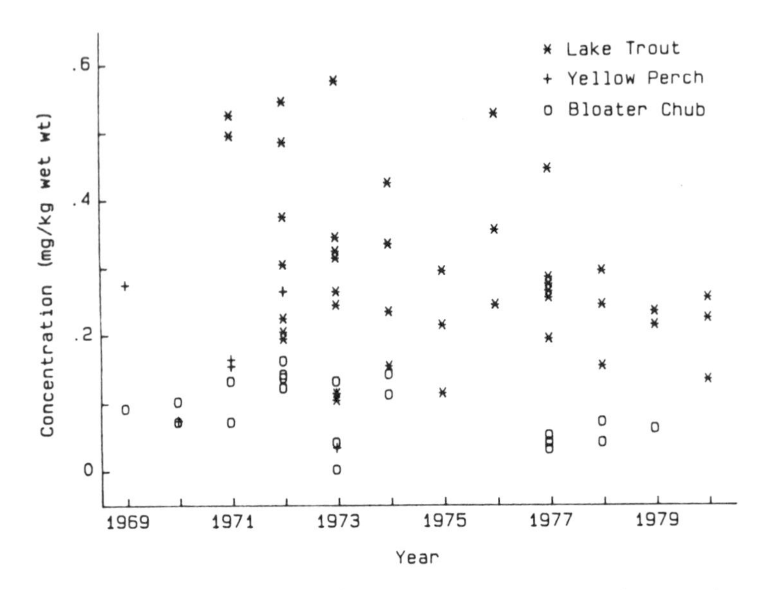

Figure 9. Mercury Residues (mg/kg wet weight) in Lake Michigan Lake Trout, Bloater Chub and Yellow Perch: 1969-1980.

LAKE HURON

Historically, Lake Huron has not been sampled with the frequency or intensity of Lake Michigan, particularly in the open waters. Figures 10 - 12 summarize the data on PCB, DDT and dieldrin contaminant burdens in whole as well as skin-on and skinless fillets. The mean annual contaminant levels are not directly comparable for Lake Huron from year to year because only one or two of the four indicator sites per lake are monitored in each year.

In 1966, an early record of organic contamination of Lake Huron fish reported DDT residues for nine fish species (Reinert 1970). In 1967, dieldrin was reported for additional Lake Huron fish (Reinert 1970), and the Bureau of Sport Fisheries and Wildlife established a National Pesticide Monitoring Program (NPMP) for several organic residues in four fish species from Saginaw Bay (Henderson et al. 1969). Since then, the NPMP has continued to monitor fish for organochlorines (Henderson et al. 1969, Henderson et al. 1971, Henderson et al. 1972, Schmitt et al. 1981, Schmitt et al. 1983, Schmitt et al. 1985). Samples were collected from Bayport, Michigan, in Saginaw Bay and offshore near Alpena, Michigan.

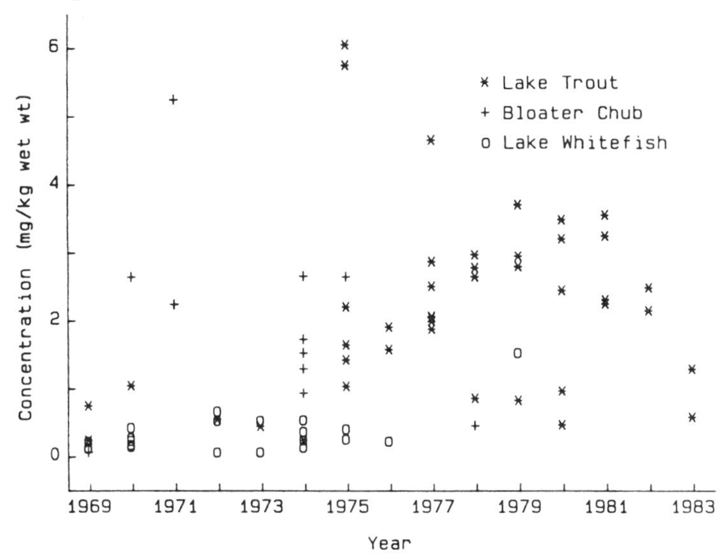

Figure 10. PCB Residues (mg/kg wet weight) in Lake Huron Lake Trout, Bloater Chub and Lake Whitefish: 1969-1983.

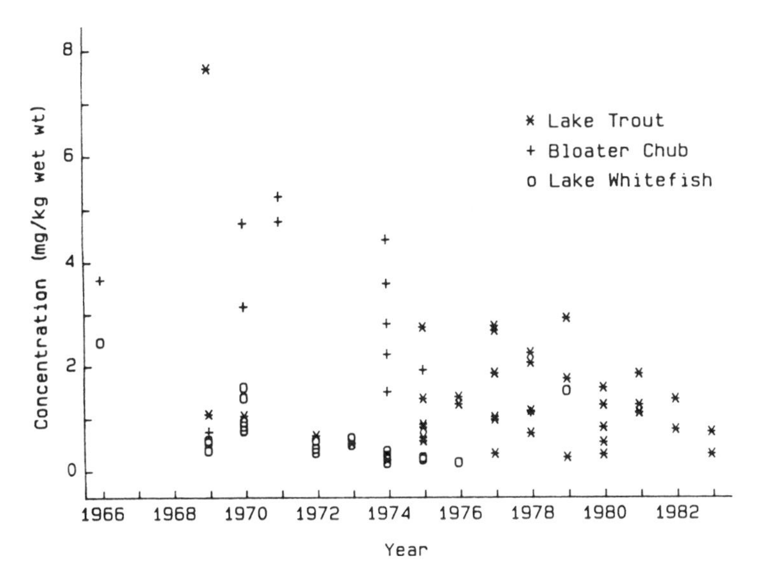

Figure 11. Total DDT Residues (mg/kg wet weight) in Lake Huron Lake Trout, Bloater Chub and Lake Whitefish: 1966-1983.

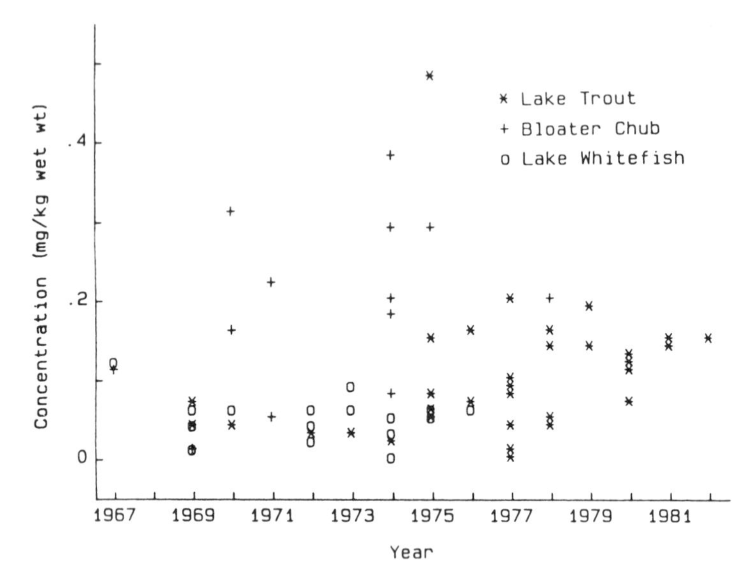

Figure 12. Dieldrin Residues (mg/kg wet weight) in Lake Michigan Lake Trout, Bloater Chub and Lake Whitefish: 1967-1982.

Concern has been greatest for the Lake Huron ecosystem with respect to the levels of PCBs in fish intended for human consumption. Although high PCB concentrations were observed generally in recent years, no individual Lake Huron fish species has shown a significantly increasing trend. In contrast, in Saginaw Bay a significantly decreasing PCB trend was observed for whole yellow perch; however, other species of fish have PCB levels significantly higher than fish from other regions of the lake.

The reason for the apparent increases in PCB levels in some Saginaw Bay fish is not known; however, one possibility is that the reduced loadings of phosphorus into the system since the early 1970s have also reduced the levels of surface active dissolved organics and particulates normally available to adsorb and precipitate the PCBs.

Most of the PCB data for lake trout were complied from 1975 through 1983 with whole fish analyses from Alpena, Michigan, and fillet analyses from Saginaw Bay. The lowest mean PCB concentration for lake trout was 0.7 mg/kg in 1969 (Frank et al. 1978) Other fish with lower PCB levels were collected in the northern basin of Georgian Bay and the open waters of Lake Huron. The highest level, 5.70 mg/kg, was found in fillet fish samples collected off Harbor Beach, Michigan, in 1975. The contaminants in lake trout from Lake Huron have generally declined since 1977; however, the declines are not statistically significant. In general, open lake fish exhibited relatively low PCB concentrations of about 1.0 mg/kg, while fillets from Saginaw Bay had the highest concentration of 3.16 mg/kg in 1981 whereas in 1980, 20 percent of the lake trout exceeded 5.0 mg/kg.

The yearly mean PCB concentrations in headless, gutted lake whitefish were relatively low and uniform between 1969 and 1976. The mean PCB levels were about 0.2 mg/kg, ranging between 0.04 and 0.6 mg/kg. The highest PCB concentration of 1.5 mg/kg was in an FDA monitoring sample in 1979.

The data are sparse for bloater chubs. The highest value was 5.2 mg/kg in 1971 for a headless, gutted bloater chub collected in Georgian Bay (Frank et al. 1978). The lowest value (headless, gutted) was 0.2 mg/kg in offshore waters in 1969. The most extensive study was conducted in 1974 (Armstrong and Lutz 1977). On the basis of whole fish analyses, PCBs were determined to be higher in Alpena than Georgian Bay and the northern and southern basins of Lake Huron. The Georgian Bay bloater chubs registered 0.89 mg/kg.

Most Lake Huron fish have contained considerably lower mean total DDT concentrations since 1973 than earlier. Since 1979-1980, many fish species have had total DDT body burdens of 1.0 mg/kg or less, and DDT contamination appears to be a declining problem. Nevertheless, occasionally the FDA limit of 5 mg/kg has been exceeded in large salmonids.

The highest mean total DDT concentration, 2.5 mg/kg, was reported for carp from Saginaw Bay in 1967. Since then the total DDT levels have decreased steadily to about 0.80 mg/kg in 1980. The ban on DDT production appears to be responsible although concentrations are still too high (Rossman 1986). The highest mean total DDT level in Saginaw Bay yellow perch was 1.66 mg/kg in 1967. Lower concentrations were reported through 1980 (about 0.15 mg/kg). The concentrations in yellow perch taken from the open waters ranged between 0.01 and 0.32 mg/kg between 1968 and 1979.

Among lake trout, the highest concentration of total DDT was 7.60 mg/kg reported for a fish taken from Georgian Bay in 1969 (Frank et al. 1978). The highest mean concentration for Saginaw Bay lake trout was 2.82

mg/kg in 1977 (Forney 1982). The lowest mean total DDT concentration was 0.49 mg/kg in lake trout taken from the main lake in 1980. Lake whitefish concentrations of total DDT showed their highest mean of 2.43 in 1966 (Reinert 1970). Concentrations essentially decreased through 1976 when the lowest mean concentration of 0.12 mg/kg was recorded.

While dieldrin has been detected in most fish taken from Lake Huron, this compound has not been consistently detected at concentrations above the FDA action level of 0.3 mg/kg. Since 1976, the dieldrin concentrations in lake trout, bloater chub and whitefish have been less than 0.20 mg/kg. Mean dieldrin concentrations are usually highest in fish collected from Alpena, Michigan, and Georgian Bay (1967-1980). The highest individual concentration of dieldrin, 0.5 mg/kg, was detected in a lake trout collected near Alpena in 1979. Other high concentrations, up to 0.38 mg/kg, were reported for bloater chub from Goderich, Ontario. As for PCB, the mean concentrations of dieldrin were highest in larger lake trout.

Mercury levels have consistently been below the FDA action level in all species except walleyes in Northern Lake Huron and Georgian Bay (Figure 13). The levels detected were: 0.55, 0.52 and 0.69 mg/kg in 1973, 1974 and 1978, respectively. High mercury concentrations were again found there in 1985, indicating possible ongoing contamination sources in these areas (Kreis and Rice 1985). Other high mercury concentrations were reported for northern pike at 0.40 mg/kg in 1974; at 0.37 mg/kg for coho salmon in 1980, lake trout in 1975 and chinook salmon in 1980; and at 0.34 mg/kg for yellow perch in 1974 and coho salmon in 1975. Walleye contained substantially greater amounts of mercury than other fish. Saginaw Bay channel catfish at about 0.3 mg/kg exhibited considerably higher mean mercury concentrations than did carp at about 0.1 mg/kg.

However, no definite conclusions can be drawn as to trends due to incomplete data.

LAKE ST. CLAIR

While high residue levels of mercury were generally found in fish from the lower Great Lakes, they appeared most frequently in samples from Lake St. Clair. The lowest levels were in fish from the open waters of the upper Great Lakes. High mercury levels also occurred

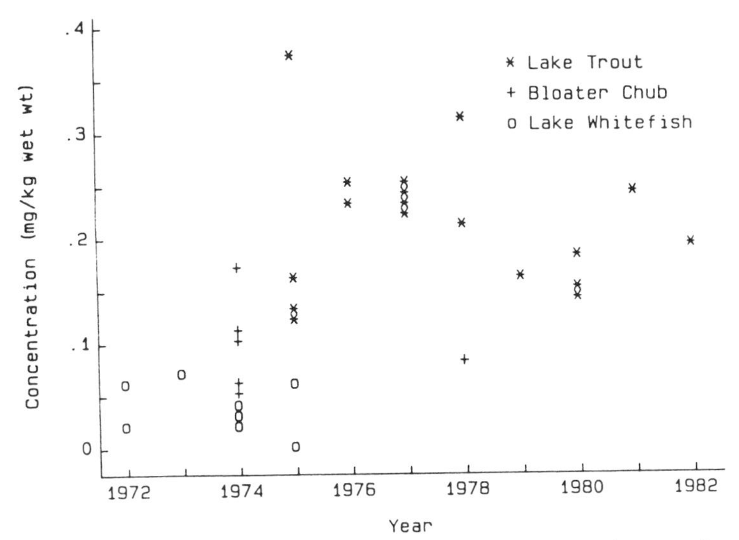

Figure 13. Mercury Residues (mg/kg wet weight) in Lake Huron Lake Trout, Bloater Chub and Lake Whitefish: 1972-1982.

much more often in some species of fish than in others. With few exceptions, the highest levels were in predatory species near the top of the food chain such as bass (Micropterus spp.), walleye, muskellunge and northern pike.

The major source of mercury to Lake St. Clair from 1950 to 1970 was a chlor alkali plant on the St. Clair River. As a result of mercury inputs from this and other lesser sources, high levels were found in many species of fish. The data show that, after the major point discharges of mercury were stopped in 1970, the levels declined dramatically for walleye (Figure 14). In 1980, only muskellunge with a mean value of 2.1 mg/kg exceeded the FDA action guideline of 1.0 mg/kg.

These rapid declines have been attributed to the cessation of industrial point discharges and the high rate of flushing in Lake St. Clair which carried the mercury contaminated sediments into the western basin of Lake Erie where they were buried. Organochlorine contaminants have generally been low in Lake St. Clair fish. DDT and dieldrin have never approached their respective FDA action levels. However, 66 percent of the channel catfish analyzed in 1972 exceeded the FDA action level for PCBs. By 1980, the decline had been great enough until only 10 percent exceeded the FDA action level (Forney 1982).

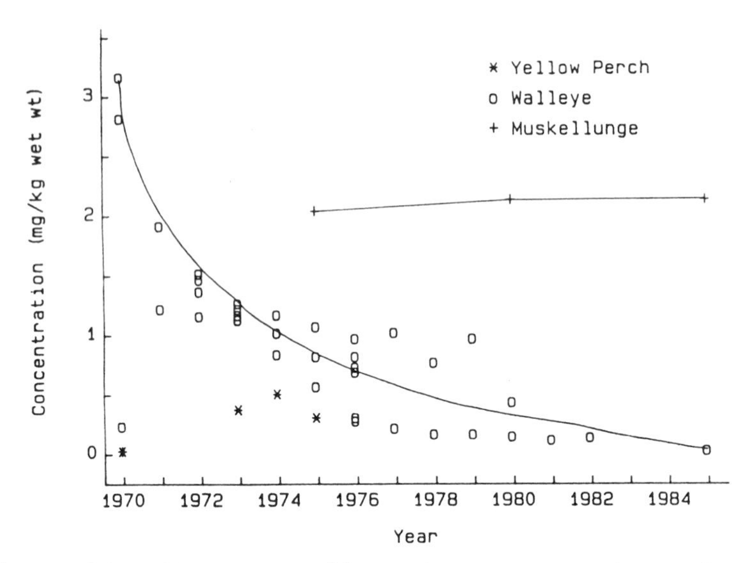

Figure 14. Mercury Residues (mg/kg wet weight) in Lake St. Clair Walleye, Yellow Perch and Muskellunge: 1970-1985.

CONCLUSION

Since 1980, the concentrations of contaminants in Great Lakes fish have declined only slightly in many cases, remained relatively constant in others, and even increased slightly in some instances. Moreover, the data often exhibit sufficient year-to-year variation that trends are not discernible. Although they are tightly regulated, some persistent compounds such as DDT, PCBs, dieldrin and mercury continue to enter the Great Lakes ecosystem and, along with deposits of the same chemicals in the sediments, especially near industrialized areas, are a source of further bioaccumulation through aquatic food chains into harvestable fish. Although the sources must now be much smaller, their environmental persistence continues and of course, the bioaccumulative capacity of the fish is ongoing. Consequently, at least traces of the contaminants have been present in fish from most of the sampling stations most of the time.

After the first regulatory actions were implemented in the early 1970s, trout and salmon samples showed an apparent dramatic decline in some of the primary contaminants such as DDT and PCBs. However, temporal trends in PCB residues were less obvious in the 1970-1974 fish samples. By 1978-79, the declines were only slight or the levels remained relatively constant. Nonetheless, the 1980 levels of chlorinated hydrocarbon contaminants in most Great Lakes fish were significantly lower than they were in 1970. However, 55 percent of the domestic fish samples tested by the FDA in 1979-80 exceeded 2.0 mg PCB/kg fresh weight; in 1980 - 81 this was 17 percent, and in 1981-82, 10 percent of the samples exceeded 2 mg/kg, including chinook salmon and their eggs and lake trout (Hoeting 1983).

Thus, despite early progress, the data clearly showed that the IJC objectives of 0.1 mg/kg for PCBs and 1.0 mg/kg for DDT in fishes in Lake Superior, for example, continued to be exceeded. Although neither of these chemicals has been in open use for more than a dozen years, only one study has shown any reliable indication of their decrease in levels of fish contamination (DeVault et al. 1985). Similarly, the concentrations of DDT residues appear to have leveled off in Lakes Superior, Huron and Michigan at about 0.30, 1.30 and 2.75 mg/kg, respectively, suggesting a continuing supply of DDT into the Great Lakes ecosystem.

The most likely current sources are from highly diffuse non-point inputs such as: agricultural and urban runoff, remobilization from the sediments, leachate from municipal and industrial landfills, municipal and industrial discharges, and illegal dumping. Probably the major loadings of PCBs to the Upper Great Lakes come from atmospheric discharges (Eisenreich et al. 1981). It is now postulated that more than 80 percent of the PCB loading to Lake Superior results from airborne deposition. The relatively greater proportions of p,p'-DDT and p,p'-DDE suggest a possible air transport from the Far East where DDT is still being used, even though the amounts were decreased in the late 1970s.

In light of the small improvement in overall pollution abatement, then, clearly the sources, fates and effects of toxic chemicals are not well enough understood presently to determine whether current laws and environmental programs are or ever will be adequate. While the concentrations of toxic chemicals in the Great Lakes are not high enough to be acutely toxic to organisms, the possible long-term subchronic effects must be considered to remain a threat because low concentrations can be raised by biomagnification and bioaccumulation. Thus, the pollution controls applied to date may not be sufficient to reduce the residue levels of these toxic chemicals to the point where human health and the aquatic ecosystem are protected. Ascertaining the extent of the problem is also complicated because the recent data often exhibit such year-to-year variations that it is not possible to discern trends. Besides the complication of the different kinds of sampling methods, the other factor is the difficulty of assessing the relative impacts of toxic chemicals from diffuse non-point sources.

The conclusion at this time must be that sufficient evidence is presently available to affirm that these chemicals may accumulate in quantities great enough, over the long if not the short term, to adversely affect the growth and reproduction of aquatic organisms and to bioaccumulate in top predator fishes to the point where they can constitute a human health threat in some circumstances. In addition, excessive contaminant residue levels may play at least a subtle role in altering the entire aquatic ecosystem. Reduced primary productivity of phytoplankton, reduced feeding rates of zooplankton, increased frequency of spinal deformities in fish, and severely reduced reproductive success of some species are among the repercussions of an increase in the total body burden of contaminants (Whittle 1985).

Besides the ongoing input of contaminants, other environmental factors also affect the levels in fish from the Great Lakes such as: the slow water turnover times, especially for the upper Great Lakes, the extended time required to achieve chemical equilibrium, continued contamination of the sediments, thefood chain and the water itself by toxic and persistent chemicals.

Finally, it is now clear that observable changes in ecosystem quality are simply slow relative to variations in chemical inputs. As the recovery of the Great Lakes from xenobiotic contamination depends on the rates of reduction in their input from external sources, the ultimate goal, of course, is the complete removal of these chemicals. Indeed, as the loading of organochlorine has decreased somewhat, the total burden has declined as well. For one thing, the removal of organochlorines from the lake water to the bottom sediments is relatively rapid; and sediments become the ultimate sink so long as the normal accumulating/burial process is faster than the rate at which the contaminants are added. Most likely, the rate of recovery of the Great Lakes relative to any further remedial action will continue to be difficult to document, but research continues in the Open Lakes Research Program originally begun by the U.S. Fish and Wildlife Service in 1969 and the Canada Department of Fisheries and Oceans in 1977. At present these are the only programs in the Great Lakes with data to document statistically significant long-term temporal and spatial trends.

REFERENCES

Armstrong, F.A.J. and A. Lutz 1977. Lake Huron, 1974: PCB, chlorinated insecticides, heavy metals and radioactivity in offshore fish. Dept. of the Environment, Fish. and Mar. Serv., Res. and Devel. Directorate, Tech. Rpt. No. 692.

Carson, R. 1962. Silent Spring. Boston: Houghton Miffin.

Clark, J.R., D. DeVault, R.J. Bowden and J.A. Weishaar 1984. Contaminant Analysis of fillets from Great Lakes coho salmon, 1980. J. Great Lakes Res. 10(1):38-47.

DeVault, D.S. 1985. Contaminants in fish from Great Lakes harbors and tributary mouths. Arch. Environ. Contam. Toxicol. 14:587-594.

DeVault, D.S. and J.A. Weishaar 1983. Contaminant analysis of 1981 fall run coho salmon (Oncorhynchus kisutch). USEPA, Great Lakes National Program Office, Chicago, IL. EPA-905/3-83-001. 16 pp.

DeVault, D.S. and J.A. Weishaar 1984. Contaminant analysis of 1982 fall run coho salmon (Oncorhynchus kisutch). USEPA, Great Lakes National Program Office, Chicago, IL. EPA-905/3-84-004. 26 pp.

DeVault, D.S., W.A. Willford and R.J. Hesselberg 1985. Contaminant trends in lake trout (Salvelinus namaycush) of the Upper Great Lakes.

USEPA, Great Lakes National Program Office, Chicago, IL. EPA 905/3-85-001. 22 pp.

Eisenreich, S.J., B.B. Looney and J.D. Thornton 1981. Airborne organic contaminants in the Great Lakes ecosystem. Environ. Sci. Technol.15(1):30-38.

Environment Canada. 1972-1977. In: 1981 Report on Great Lakes Water Quality. Appendix. Great Lakes Surveillance. Great Lakes Water Quality Board Report to the International Joint Commission.

Eschmeyer, R.H. and A.M Phillips, Jr. 1965. Fat content of the flesh of siscowets and lake trout from Lake Superior. Trans. Amer. Fish. Soc. 94(1):62-74.

Fisheries and Oceans Canada 1977-1980. In: 1981 Report on Great Lakes Water Quality. Appendix. Great Lakes Surveillance. Great Lakes Water Quality Board Report to the International Joint Commission.

Forney, J. 1982. Great Lakes Environmental Contaminants Survey. Summary Report. 1972-1980. Michigan Department of Natural Resources, Office of Toxic Materials Control, Publication. No. 3730-0038, Lansing, MI 48909.

Frank, R., M. Holdrinet, H.E. Braun, D.P. Dodge and G.E. Spangler 1978. Residues of organochlorine insecticides and polychlorinated biphenyls in fish from Lakes Huron and Superior, Canada - 1968-1976. Pestic. Monit.J. 12:60-68.

Frank, R., R.L. Thomas, M. Holdrinet, A.L.W. Kemp and H.E. Braun 1979. Organochlorine insecticides and PCB in surficial sediments (1968) and sediment cores (1976) from Lake Ontario. J. Great Lakes Res. 5:18-27.

Frank, R., R.L. Thomas, H.E. Bralin, D.L. Gross and T.T. Davies 1981. Organochlorine insecticides and PCB in surficial sediments of Lake Michigan (1975). J. Great Lakes Res. 7(1):42-50.

GLNPO 1981. A strategy for fish contaminant monitoring in the Great Lakes. USEPA, Great Lakes National Program Office, Chicago, IL 60605.

Hartig, J.H. and M.E. Stifler 1980. Water quality and pollution control in Michigan, 1980. Michigan Department of Natural Resources, Publication No. 4833-9803, Lansing, MI 48909.

Henderson, C., W.L. Johnson and A. Inglis 1969. Organochlorine insecticide residues in fish--National Pesticide Monitoring Program. Pestic. Monit. J. 3(3):145-171.

Henderson, C., A. Inglis and W.L. Johnson 1971. Organochlorine insecticide residues in fish, fall 1969--National Pesticide Monitoring Program. Pestic. Monit. J. 5(1):1-11.

Henderson, C., A. Inglis and W.L. Johnson 1972. Mercury residues in fish, 1969-1970--National Pesticide Monitoring Program. Pestic. Monit. J. 6(3):144-159.

Hoeting, A.L. 1983. FDA regulation of PCB in food. In: PCBs: Human and Environmental Hazards, F.M. D'Itri and M.A. Kamrin (eds.), Butterworth Publishing, Woburn, MA, pp. 393-407.

IJC 1977. The Waters of Lake Huron and Lake Superior Vol. III (Part B), Lake Superior. Report to the International Joint Commission by the Upper Lakes Reference Group, Windsor, Ontario, 574 pp.

IJC 1978a. Great Lakes Water Quality Status Report, Appendix E. Great Lakes Water Quality Board, International Joint Commission, Windsor, Ontario, Canada.

IJC 1978b. Status Report on Organic and Heavy Metal Contaminants in the Lakes Erie, Michigan, Huron and Superior Basins. International Joint Commission, Great Lakes Water Quality Board, Appendix E, Windsor,Ontario, 373 pp.

IJC 1981. Report on Great Lakes Water Quality. International Joint Commission, Great Lakes Water Quality Board, Appendix: Great Lakes Surveillance, 174 pp.

IJC 1983. An inventory of chemical substances identified in the Great Lakes ecosystem. Report to the Great Lakes Water Quality Board, International Joint Commission, Windsor, Ontario, 6 volumes.

Inglis, A., C. Henderson and W.L. Johnson 1971. Expanded program for pesticide monitoring of fish. Pestic. Monit. J. 5:47-49.

Jaffe, R., E.A. Stemmler, B.D. Eitzer and R.A. Hites 1985. Anthropogenic, polyhalogenated, organic compounds in sedentary fish from Lake Huron and Lake Superior tributaries and embayments. J. Great Lakes Res. 11(2):156-162.

Jensen, A.L., S.A. Spigarelli and M.M. Thommes 1982. PCB uptake by five species of fish in Lake Michigan, Green Bay of Lake Michigan, and Cayuga Lake, New York. Can. J. Fish. Aquat. Sci. 39:700-709.

Kreis, R.G. Jr. and C.P. Rice 1985. Status of organic contaminants in Lake Huron: atmosphere, water, algae, fish, herring gull eggs, and sediment. Great Lakes Research Division Special Report No. 114, The University of Michigan, Ann Arbor, Michigan 48109. 169 pp. + appendix.

Kuehl, D.W., E.N. Leonard, B.C. Butterworth and K.L. Johnson 1983. Polychlorinated chemical residues in fish from major watersheds near the Great Lakes, 1979. Environ. Int. 9:293-299.

Lauer, T. 1980. Results of cooperative fish contaminant monitoring program between the Indiana State Board of Health and the Indiana Department of Natural Resources. ISBH, Indianapolis, IN.

Lowe, T.P., T.W. May, W.G. Brumbaugh and D.A. Kane 1985. Concentrations of seven elements in freshwater fish, 1978-1981. National Contaminant Biomonitoring Program, Arch. Environ. Contam. Toxicol. 14:363-388.

May, T.W. and G.L. McKinney 1981. Cadmium, lead, mercury, arsenic and selenium concentrations in freshwater fish, 1976-1977. National Pesticide Monitoring Program. Pestic. Monit. J. 15(1):14-38.

MDA 1975a. Great Lakes Environment Contaminants Survey. Michigan Department of Agriculture, Lansing, MI 48909.

MDA 1975b. Summary of 1974-1975 GLECS Data - Lake Huron. Unpublished data, Michigan Department of Agriculture, Lansing, MI 48909.

MDA 1976. Summary of 1975 GLECS Data - Lake Huron. Unpublished data, Michigan Department of Agriculture, Lansing, MI 48909.

MDNR 1979. Great Lakes Environmental Contaminants Survey 1976-1978. Michigan Department of Natural Resources, Office of Toxic Material Control, Lansing, MI 48909.

MWRC 1975. Limnological surveys of Lakes Superior and Huron nearshore waters, 1974-1975. Michigan Water Resources Commission, Bureau of Water Management, Lansing, MI. EPA Contract No. R00514601, ULRG Project D-16.

Murphy, T.J. and C.P. Rzeszutko 1977. Precipitation inputs of PCBs to Lake Michigan. J. Great Lakes Res. 3:305-312.

Neidermyer, W.J. and J.J. Hickey 1976. Chronology of organochlorine compounds in Lake Michigan fish, 1929-1966. Pestic. Monit. J. 10:92-95.

ODNR 1980. 1979 Status of PCBs in Lake Erie Fishes. Ohio Department of Natural Resources, Columbus, OH.

OMOE 1976. Georgian Bay - Lake Huron - St. Marys River Fish Analysis. Unpublished report, Ontario Ministry of the Environment, Toronto, Ontario. ULRG Project D-23.

OMOE 1977. The decline in mercury concentration in fish from Lake St. Clair, 1970-1976. Ontario Ministry of the Environment, Report No.A)S77-3, 85 pp.

Olsson, M., S. Jensen and L. Reutergard 1978. Seasonal variation of PCB levels in fish - An important factor in planning aquatic monitoring programs. Ambio. 7(2):66-69.

O'Shea, T.J. and J.L. Ludke 1979. Monitoring fish and wildlife for environmental pollutants. U.S. Fish Wildl. Serv., 12 pp., Available from U.S. Govt. Printing Office, Washington, DC, as Stock No. 024-010-00512-2.

Parejko, R., R. Johnston and R. Keller 1975. Chlorohydrocarbons in Lake Superior lake trout (Salvelinus namaycush). Bull. Environ. Contamin. Toxicol. 14: 480-488.

Poff R.J. and P.E. Degurse 1970. Wisconsin Department of Natural Resources Report 34.

Powers, R.A. 1976. A review of PCB's. Michigan Water Resource Commission, Department of Natural Resources, Lansing, Michigan 48909.

Reinert, R.E. 1970. Pesticide concentrations in Great Lakes fish. Pestic. Monit. J. 3(4):233-240.

Reinke, J., J.F. Uthe and D. Jamieson 1972. Organochlorine pesticide residues in commercially caught fish in Canada - 1970. Pestic. Monit. J. 6(1):43-49.

Rockwell, D.C., D.S. DeVault III, M.F. Palmer, C.V. Marion, and R.J Bowden 1980. Lake Michigan Intensive Survey 1976-1977. Great Lakes National Program Office, U.S. Environmental Protection Agency, 536 South Clark Street, Chicago, IL 60605, Report No. EPA-905/4-80-003-A.

Rodgers, P.W. and W.R. Swain 1983. PCB loading of trends in Lake Michigan. J. Great Lakes Res. 9(4): 548-558.

Rohrer, T.K., J.H. Hartig and J.C. Forney 1981. Organochlorine and heavy metal residues in coho and chinook salmon of the Great Lakes - 1980. MDNR Publ. No. 3730-0031. Michigan Department of Natural Resources, 23 pp.

Rohrer, T.K., J.C. Forney and J.H. Hartig 1982. Organochlorine and heavy metal residues in standard fillets of coho and chinook salmon of the Great Lakes - 1980. J. Great Lakes Res. 8:623-634.

Rossman, R. 1986. Lake Huron 1980 intensive surveillance: management and summary, Great Lakes Research Division, Special Report No. 118, The University of Michigan, Ann Arbor, Michigan 48109. 59 pp. + appendix.

Rybicki, R.W. and M. Keller 1978. The Lake Trout Resource in Michigan Waters of Lake Michigan, 1970-76. Michigan Department of Natural Resources, Fisheries Research Report No. 1863.

Safe, S. 1984. Polychlorinated biphenyls (PCBs) and polybrominated biphenyls (PBBs): biochemistry, toxicology, and mechanism of action. CRC Crit. Rev. Toxicol. 13:319-393.

Schmitt, C.J. J.L. Ludke and D.F. Walsh 1981. Organochlorine residues in fish: National Pesticide Monitoring Program, 1970-1974. Pestic. Monit. J. 14:136-206.

Schmitt, C.J., M.A. Ribick, J.L. Ludke and T.W. May 1983. Organochlorine residues in freshwater fish, 1976-79. National Pesticide Monitoring Program. U.S. Fish Wild. Serv. Resour. Publ. 152, Washington, D.C. 62 pp.

Schmitt, C.J., J.L. Zajicek and M.A. Ribick 1985. Residues of organochlorine chemicals in freshwater fish, 1980-81. National Pesticide Monitoring Program. Arch. Environ. Contam. Toxicol. 14:225-260.

Sheffy, T.B. 1977. Status report on the PCB problem in Wisconsin. Wisconsin Department of Natural Resources, Madison, WI.

Sheffy, T.B. and J.R. St. Amant 1980. Toxic substances survey of Lakes Michigan, Superior, and tributary streams; first annual report. Wisconsin Department of Natural Resources, Madison, WI.

Simmons, M.S. 1984. PCB contamination in the Great Lakes. In: Toxic Contamination in the Great Lakes, J.O. Nriagu and M.S. Simmons (eds.), John Wiley, New York, pp. 287-309.

Strachan, W.M.J. and G.E. Glass 1978. Organochlorine substances in Lake Superior. J. Great Lakes Res. 4(3-4):389-397.

Swain, W.R. 1978. Chlorinated organic residues in fish, water and precipitation from the vicinity of Isle Royale, Lake Superior. J. Great Lakes 4:398-407.

Veith, G.D. 1975. Baseline concentrations of polychlorinated biphenyls and DDT in Lake Michigan fish, 1971. Pestic. Monit. J. 9:21-29.

Veith, G.D., D.W. Kuehl, F.A. Puglisi, G.E. Glass and J.G. Eaton 1977. Residues of PCBs and DDT in the Western Lake Superior ecosystems. Arch. Environ. Contamin. Toxicol. 5:587-599.

Veith, G.D., D.W. Kuehl, E.N. Leonard, K. Welch and G. Pratt 1981. Polychlorinated biphenyls and other organic chemical residues in fish from major watersheds near the Great Lakes, 1978. Pestic. Monit. J. 5:1-8.

Veith, G.D., D.W. Kuehl, E.N. Leonard, F.A. Puglisi and A.E. Lemke 1979. Polychlorinated biphenyls and other organic chemical residues in fish from major watersheds of the United States, 1976. Pestic. Monit. J. 13:1-11.

Veith, G.D. and G.F. Lee 1971. PCBs in Fish from the Milwaukee region. Proc. 14th Conference of Great Lakes Research, International Association of Great Lake Research, pp. 157-169.

WDNR 1985. Summary of contaminants in Lake Michigan fish, unpublished data. Wisconsin Department of Natural Resources. Madison, WI 53707.

Walsh, D.F., B.L. Berger and J.R. Bean 1977. Mercury, arsenic, lead, cadmium and selenium residues in fish 1971-1974 - National Pesticide Monitoring Program. Pestic. Monit. J. 11:5-34.

Whittle, D.M. 1985. Trends in contaminant burdens of the Lake Superior fish community and its forage base. Department of Fisheries and Oceans,Great Lakes Fisheries Research Branch, 867 Lakeshore Road, Burlington, Ontario L7R-4A6, Unpublished manuscript.

Willford, W.A., R.J. Hesselberg and L.W. Nicholson 1976. Trends of polychlorinated biphenyls in three Lake Michigan fishes. In: Proceedings of the National Conference on Polychlorinated Biphenyls. U.S. Environmental Protection Agency, Office of Toxic Substances, Washington, D.C. EPA-560/6-75-004. pp. 177-181.

Williams, D.J. 1975. A review of published and other data on residues of DDT, dieldrin, mercury and PCBs in fish in the Great Lakes. Draft document prepared for Great Lakes Biolimnology Laboratory, Department of Environment.

Wszolek, P.C., D.J. Lisk, T. Wachs and W.D. Young 1979. Persistence of polychlorinated biphenyls and 1,1-dichloro-2,2-bis(p-chlorophenyl) ethylene (p,p'-DDE) with age in lake trout after 8 years. Environ. Sci. Technol. 13(1):1269-1271.

Young, W.D., W.H. Gutenmann and D.J List 1972. Residues of DDT in lake trout as a function of age. Environ. Sci. Technol. 6(5):451-452.

CHAPTER 5

UNITING HABITAT QUALITY AND FISHERY PROGRAMS IN THE GREAT LAKES

Carlos M. Fetterolf, Jr.

Great Lakes Fishery Commission, 1451 Green Road, Ann Arbor, Michigan 48105 USA

ABSTRACT

Despite the remarkable recoveries of Great Lakes fisheries since the 1950s and water quality since the 1970s many of today's fishery problems are related to environmental conditions. Recognizing the public's resource must receive full consideration in the present and proposed uses of the lakes, in 1981 the 12 agencies with mandated responsibility for the welfare of Great Lakes fisheries developed and signed a Joint Strategic Plan for management of Great Lakes fisheries under the aegis of the Great Lakes Fishery Commission (GLFC). The four fundamental strategies involve consensus, accountability, environmental management and management information. Four strategic procedures of the plan were designed to aid in coordinating environmental and fishery agency management efforts into a complementary process:

1. referral of environmental issues which impede achievement of fishery objectives to the GLFC,

2. representation by the GLFC in such issues to the most appropriate body,

3. identification of fishery agency plans to achieve environmental objectives, and

4. establishment of a Habitat Advisory Board to assist lake committees and the GLFC in environmental issues.

In 1986 the Habitat Advisory Board added to the Strategic Plan a component for the protection, rehabilitation and enhancement of habitat required to ensure accomplishment of fishery management objectives. The fish habitat component emphasizes five areas: information, legal, working arrangements, intervention/advocacy, and public participation. Fish are of central interest to the remediation and rehabilitation initiatives underway throughout the Great Lakes. For fish habitat management and planning to be successful, a web of involvement must be developed at all levels of responsibility across agency and jurisdictional boundaries. The habitat needs for fish communities must be decided by fishery managers, and consensus on program objectives needs to be developed among the many related agencies. Optimum results can only be achieved by uniting habitat quality and fishery programs in both the planning and operational stages.

INTRODUCTION

Management of the Great Lakes is especially complicated because the lakes are shared between Canada and the United States, eight states and the province of Ontario. Many jurisdictions and agencies have responsibility for different aspects of management. This complexity has resulted in a series of efforts to develop effective, cooperative agreements to resolve transboundary, interstate, and international ecosystem issues. The most important products have been the Boundary Waters Treaty, 1909; the Convention on Great Lakes Fisheries, 1956; the Great Lakes Basin Compact, 1968; the Great Lakes Water Quality Agreement, 1972 and 1978; the Joint Strategic Plan for Management of Great Lakes Fisheries, 1981; and the Great Lakes Charter, 1985. This discussion will focus on those instruments most involved with habitat quality and fisheries.

BOUNDARY WATERS TREATY, THE INTERNATIONAL JOINT COMMISSION AND THE CANADA/UNITED STATES WATER QUALITY AGREEMENT

The Boundary Waters Treaty of 1909 was created by Canada and the U.S. to prevent disputes relating to boundary water usage and to settle questions arising along the entire boundary from coast to coast (IJC 1965). The Treaty created the International Joint Commission (IJC) which has provided the framework for cooperation on questions relating to water and air pollution and the regulation of water levels and flows. In these arenas the IJC has responsibilities for judging certain applications, conducting investigations, making non-binding recommendations to governments, monitoring compliance and coordinating implementation of recommendations accepted by the governments.

The IJC has specific Great Lakes responsibilities under the Great Lakes Agreement of 1972 (amended in 1978) (IJC 1978). Under the Agreement the purpose of the Parties is to:

> ... "restore and maintain the chemical, physical and biological integrity of the waters of the Great Lakes Basin Ecosystem."...

To achieve this purposes, the Parties agreed to "make a maximum effort to develop programs, practices and technology necessary for a better understanding of the Great Lakes Basin ecosystem and to eliminate or reduce to the maximum extent practicable the discharge of pollutants into the Great Lakes System." The major responsibility for ecosystem quality management rests with the federal, state and provincial environmental protection agencies, or with the environmental bureaus of the traditional natural resource agencies. The goals and objectives are stated in the Agreement and its annexes which focus on specific issues. The Great Lakes Regional Office of IJC monitors compliance with the Agreement. Working with two boards and a plethora of committees, the IJC attempts to carry out the charges in the Agreement. Much habitat improvement is evident, particularly from reduced nutrient loading in many embayments, but many fishery habitat quality goals have not been achieved. However, fishery managers have not necessarily been clear in expressing the habitat needs of the fishery resource over the past century. The official responsibility for environmental quality management often does not rest with the fishery manager's agency, and when the environmental movement started to have an impact in the mid-60s, fishery people tended to depend on the environmental quality expert.

CONVENTION ON GREAT LAKES FISHERIES

Development of the Laurentian Great Lakes Basin since the early 1800s has had a major, continuing influence on the Great Lakes ecosystem and its fish populations. A wide range of activities including lumbering, farming, hydropower development, municipal and industrial waste disposal, canal construction, invasion by marine species, harbor and building site development, stocking of both exotic and native fish species, and intensive, species-selective fishing contributes to the many changes man has initiated in what was formerly a simple, slowly evolving flora and fauna (Berst and Spangler 1973, Christie 1973, Hartman 1973, Lawrie and Rahrer 1973, Smith 1972, Wells and McLain 1973). In the world renowned lake whitefish (*Coregonus clupeaformis*) and lake trout (*Salvelinus namaycush*) fisheries of Lakes Huron, Michigan and Superior, fishing effort seemed to balance productivity for a hundred years while the fisheries for other species experienced irregular fluctuations. In Lakes Erie and Ontario there have been concerns for unstable, declining fisheries since the mid-1800s. (Fetterolf 1980). Atlantic salmon were lost from Lake Ontario in the 1890s and several species from Lake Erie in the 1900s.

The sea lamprey (*Petromyzon marinus*) had a great impact on the fisheries of the Great Lakes, especially the upper three lakes (Pearce et al. 1980, Smith and Tibbles 1980). Native to the Atlantic Ocean, the sea lamprey feeds by attaching to other fish with its suctorial mouth and extracting body fluids. A sea lamprey may kill as much as 40 pounds of fish in its adult life. Sea lampreys probably entered Lake Ontario via the Hudson River and its extension, the Erie Canal, opened to Lake Ontario in 1819. It gained access to lake Erie via the Welland Canal around Niagara Falls in 1929 and spread to Lake Superior by 1946 (Lamsa et al. 1980). Through a combination of heavy fishing pressure and sea lamprey predation, the lake trout commercial catch from Lakes Michigan and Huron, which had averaged 13 million pounds for many years, dropped to zero (Baldin et al. 1979).

Efforts to establish international fishery commissions and/or effective complementary regulations and management programs for Great Lakes fisheries failed repeatedly from 1893 to 1952 (Fetterolf, 1980). However, the spread of the sea lamprey to the upper lakes was recognized as an impending international catastrophe for the fisheries. This threat provided an added incentive to recast and complete earlier negotiations, and the Convention on Great Lakes Fisheries was ratified in 1955 by Canada and the United States (GLFC 1983). The Convention focused on determining "the need for and the type of measures which will make possible the maximum sustained productivity in Great Lakes fisheries of common concern."

GREAT LAKES FISHERY COMMISSION

The Convention established the Great Lakes Fishery Commission to formulate fishery programs, determine the best management measures, coordinate and undertake research, make recommendations to and advise the governments, to publish on ways and means to control the sea lamprey. Only by working closely with the state, provincial and federal fishing agencies could this be accomplished.

The Commission (Figure 1) carries out its responsibilities for sea lamprey control and research through its contract agents, the U.S. Fish and Wildlife Service and Fisheries and Oceans Canada, and pursues much of its program through a committee structure involving representatives of the academic community and agencies with mandates for fishery management and research.

Managers have made much progress since 1955 in rehabilitating Great Lakes fish communities and habitat. Sea lamprey and overabundant forage species such as the alewife (<u>Alosa pseudoharengus</u>) are largely under control. Stocking of lake trout and Pacific salmon (<u>Oncorhynchus</u> sp.) plus catch restrictions have restored predator populations. From a fishery which can only be described as devastated, the annual regional economic impact of commercial fisheries has recently been estimated at $270 million and that of the recreational fisheries from $2.0 - $4.0 billion (Talhelm 1986a). Some 55 million angler days are spent in pursuit of Great Lakes fish every year (Talhelm 1986b.)

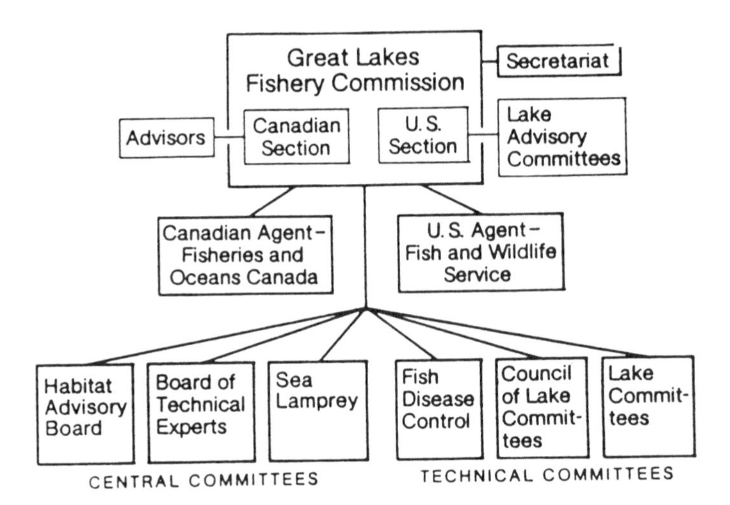

Figure 1. Organization chart of the Great Lakes Fishery Commission.

ENVIRONMENTAL PROBLEMS

Some 37 million people live in the Basin and 24 million drink Great Lakes water. With people come municipal sewage, industry and toxic wastes. The lakes responded vigorously to exotic fish invaders, overfishing and pollution. We had filamentous algae in nuisance quantities on many beaches. We had dramatic fish species changes and dieoffs. We continue to have difficulty getting lake trout to reproduce (Eshenroder et al. 1984, Fetterolf 1985), and we have the problem of chemical contaminants in fish. The nest of every Great Lakes fisherman and fisheries manager has been fouled by the point source, land runoff, waste dump seepage and atmospheric inputs of toxic chemicals. Public health advisories are issued by every state and Ontario warning against consumption of certain species and sizes of Great Lakes fish. While most fish from most areas are acceptable, the residue levels in some large lake trout, chinook salmon (*Oncorhynchus tshawytscha*), brown trout (*Salmo trutta*), and carp (*Cyprinus carpio*) exceed acceptable risk levels if the fish are eaten in large quantities over extended periods.

Chemical residues in Great Lakes fish cast a pall over the social and economic aspects of Great Lakes fisheries. They have created a very real problem for commercial fishermen, processors, and retailers; a shadow of doubt in the minds of every consumer and recreational fisherman; an added question for the fishery manager; a symbol of defeat for the pollution control agencies; and a mark for every environmental management critic to flaunt as an example of the failure of the regulatory system. Contaminants deny full use of the Great Lakes fishery resource. For example, American eels (Anguilla rostrata) were recently the most valuable commercial fish in Lake Ontario at $2 - $3 per pound but can no longer be sold because of Mirex and PCB residues.

Despite many years of expensive waste management efforts, the IJC recently identified 42 areas of concern where water quality objectives are not met. Fish and aquatic life are often the most sensitive indicators of inadequate habitat quality. It can be generally stated that fishery resources would benefit if water quality improved in these 42 areas of concern. Each of these areas are nearshore habitats such as connecting channels, embayments and the lower reaches of tributaries, the areas where the majority of Great Lakes fishing occurs. These nearshore areas encompass critical habitat for many species of Great Lakes fish for migration, spawning and development of young; they are most vulnerable to point source discharges, tributary loadings and runoff; their condition influences offshore ecosystem health and productivity; and there is great potential for further development of fisheries in these nearshore areas.

Fishery scientist were not actively involved in the process when decisions were made which resulted in contaminant residues in fish, or in the waste management decisions which created 42 areas of concern on the Great Lakes. How can fishery people assure that fishery interests will be at the table when the high stakes games are played or when the plans, agreements and decisions on environmental matters are made? Only by planning to be there and then making damn sure we are. How can this representation come about?

A POLICY OF FISHERY REPRESENTATION

In 1980 the Great Lakes Fishery Commission developed a policy advocating fishery representation on advisory and decision-making groups involved with environmental concerns. However, you can not force fishery managers to get involved with these groups. They have to want to get involved and many do. Many others have been lulled into thinking that only environmental managers have the responsibility and the ability to plan for water and habitat quality. For some reason, fishery managers have concluded that environmental mangers (aquatic biologists, limnologists, chemists, bacteriologists, engineers and lawyers) have the fishery resource's best interests at heart when they manage with water quality standards, best practical treatment, best available treatment, etc. Fishery managers must be involved to look out for fishery interests. So, next came a plan from fishery managers to be more of an insider than an outsider.

THE JOINT STRATEGIC PLAN FOR MANAGEMENT OF GREAT LAKES FISHERY RESOURCES

The Commission's Council of Lake Committees, made up of senior fishery executives from each state and provincial agency managing the Great Lakes fishery, (in 1986 the Chippewa-Ottawa Treaty Fishery Management Authority and the Great Lakes Indian Fish and Wildlife Commission were included in the Council of Lake Committees) recognizing that threats to the fishery resource and opportunities for managing the fishery require greater capability than any one agency or government can provide, recommended that the Fishery Commission develop a strategic plan for management of Great Lakes fisheries. As in so much of its work, the Commission agreed to facilitate the joint efforts of its cooperators by providing guidance at the policy level and a neutral, resource-oriented forum in which mutually beneficial programs could be developed. The Commission secured the commitment of 12 agency directors for development of the plan and constituted them into a Committee of the Whole with veto power. Two years later in 1981, the agencies signed their own Joint Strategic Plan for Management of Great Lakes Fisheries (Can. Dept. Fish. and Oceans et al. 1980). The acronym is SGLEFMP (siggle-fump).

The plan strives "to secure fish communities, based on foundations of stable self-sustaining stocks, supplemented by judicious plantings of hatchery-reared fish, and to provide from these communities an optimum contribution of fish, fishing opportunities and associated benefits to meet needs identified by society for: wholesome food, recreation, employment and income, and a healthy human environment."

The plan provides four strategies for dealing with the issues and achieving the goals. The first is

consensus. "Consensus must be achieved when management will significantly influence the interests of more than one jurisdiction."

A second strategy is **accountability.** "Fishery management agencies must be openly accountable for their performance." Each agency will keep others informed of their programs, operational objectives, targets and performance.

The third strategy is **environmental management.** "Fishery agencies shall endeavor to obtain full consideration by the Great Lakes environmental management agencies of the potential impacts of their activities and decisions on fishery needs and objective." Many current fishery problems are environmental quality problems, but fishery agencies often lack jurisdiction and adequate influence over environmental management decisions. This strategy encourages the fishery agencies to work with the environmental agencies to identify the impacts of environmental actions on the fishery resource.

The fourth strategy is **management information.** "Fishery agencies must cooperatively develop means of measuring and predicting the effects of fishery and environmental management decisions."

The plan lays out 13 strategic procedures for lake committees, fishery agencies and the Commission. Using this plan for guidance, the agencies and Commission can ensure that the public's fishery resource receives full consideration from other uses of our Great Lakes.

Four of the strategies deal directly with environmental concerns:

1. The lake committees will identify environmental issues which may impede achievement of their fishery objectives and refer these to the Great Lakes Fishery Commission.

2. Unresolved environmental issues may be referred by lake committees to the Great Lakes Fishery Commission which will represent fishery interests in these issues to the most appropriate body (e.g. International Joint Commission, United States Environmental Protection Agency, United States State Department, Canada Department of External Affairs).

3. Each fishery agency should identify its plans for achieving the fish community and environmental objectives identified by the lake committees noting proposed collaboration with environmental and other agencies as well as its own proposed activities.

4. The Great Lakes Fishery Commission will create an expert Habitat Advisory Board (HAB) and charge it to assist each lake committee to develop environmental objectives essential to achieve its fishery objectives.

GUIDELINES FOR FISH HABITAT MANAGEMENT AND PLANNING

The Habitat Advisory Board became active in 1984 and was charged to advise the Commission on a wide range of fishery-related habitat and water quality and quantity issues. HAB assists the Commission in its advocacy role for fishery resources and as a catalyst for the development of improved habitat assessment and management capabilities.

In May 1986 the Fishery Commission received HAB's Guidelines for Fish Habitat Management and Planning in the Great Lakes to protect, rehabilitate and enhance habitat required to ensure accomplishment of fishery management objectives. Anticipating the Commissions' endorsement of this or a slightly modified plan, the HAB will work with the Lake Committees toward more involvement by fishery managers in habitat matters. Major impetus was provided by the annual meeting of the Fishery Commission when the directors and ministers of the Great Lakes natural resource agencies (Committee of the Whole) encouraged more active involvement by fishery personnel in environmental issues. In other words, the fishery managers' bosses told them to become a bigger influence on the habitat management front. HAB's concept of fish habitat includes all the physical and chemical factors and many of the biological ones on which the fish are dependent. Fishery managers should never forget that habitat quality, if reversible, should not constrain fishery objectives. Think big! For example, get the Corps of Engineers working with the fishery team!

HAB's guidelines stress that planning processes should be harmonized both vertically (from field to policy levels) and horizontally (across the various mandates which influence fisheries). Decision makers at all levels should participate in development and implementation in appropriate ways. The key to successful implementation of the planning process is an effective set of arrangements for influencing decisions in policy fields beyond line control of fishery managers.

To be effective, the habitat component of fishery management plans must be strong in five areas: best use of scientific data, strong socio-economic arguments, establishment of a strong basis in legal authorities, adequate emphasis on institutional arrangements and decision-making processes including intervention/advocacy; and support and participation by clients and public.

Information

Information should be applied in systematic ways to examine choices and make decisions. Critical attention must be paid to assumptions employed by decision makers. The establishment of habitat objectives by comparing requirements of desired fish communities with current habitat conditions will be a rigorous test of our habitat assessment programs and ecological knowledge.

Socio-Economic

Fishery objectives need to be supported by strong broadly defined socio-economic arguments. This will build a broad constituency and fit with most management goals.

Legal

Habitat planning must be carried out pragmatically under a broad umbrella of legal authorities, many outside the jurisdiction of fishery agencies. The legal basis becomes a set of explicit statements on how each jurisdiction will use specified statutes for specific purposes. Fishery agencies should use full legal authority available to them, but strong socio-economic arguments will also be needed. The Boundary Waters Treaty and Canada-U.S. Water Quality Agreement could be more effectively used in the interest of fishery management.

Working Arrangements - Intervention/Advocacy

Well-defined arrangements among fishery agencies and with other agencies form the ongoing machinery for making decisions in the best interests of fisheries. Fisheries plans will provide a factual basis for intervention in habitat issues and supportive evidence for positions taken by fishery agencies. Commitments to intervention/advocacy and appropriate protocols for its use should be specified. Similarly, working arrangements among all institutions should be specified. It is implicit that wildlife habitat concerns will be included whenever appropriate.

Public Participation

Plans should incorporate policies and programs to strengthen public support and political will and provide a better coordinated, stronger voice for fisheries. Habitat information should be in fishery terms meaningful to the public. Plans should pay adequate attention to building constituencies for fishery agencies and the GLFC. Therefore, public participation should be early and meaningful.

RECOMMENDATIONS

The following recommendations, aimed at various targets from individuals to lake committees, to agency heads and to the GLFC itself, are intended to stimulate broadly based support for the habitat components of fishery plans. Fishery benefits can be optimized when dealing with habitat concerns by:

A: Developing a web of involvement with client groups and the decision makers,

B: Involving fishery scientists in the development of habitat objectives,

C: Uniting the habitat quality programs with fishery programs, and

D: Achieving consensus among the management groups, the user groups and the clients.

Habitat management and planning guidelines are part of the ecosystem approach to management where all the involved groups share the responsibilities for the quality of the Great Lakes ecosystem. The goal is to leave to future generations a management legacy based on preventive rather than remedial measures. We can do this by managing the Great Lakes Basin as a home (Christie et al. 1986).

CONCLUSION

Let us consider the words of Henry David Thoreau:

> ... "A lake is the landscape's most beautiful and expressive feature. It is the earth's eye; looking into which the beholder measures the depth of his own nature."...

A few years ago, as beholders looking into the Great Lakes as the earth's eye, Canadians and Americans had to share the shame because of our exploitive nature and ignorance as to how we should care for this ecosystem.

Today we can even have a touch of pride in how we are managing the lakes. What do we use for a yardstick in measuring our success? Just ask the people around the lakes a few questions:

- -How's the fishing?
- -Do the fish taste good?
- -Are there any warnings about their consumption?
- -How does the drinking water taste?
- -Are there any warnings about drinking it?
- -Is the beach clean of algae and floating wastes?
- -Are there warnings about swimming in the water?
- -Do we have easy access to the lakes?
- -Do the water birds, fur bearers and fish have healthy populations?
- -Is it a pleasantly aesthetic experience to visit the lakes?

If we can answer yes to all of those, then we are on our way to successful management of water quality, fisheries and wildlife.

ACKNOWLEDGEMENTS

Andy Lawrie, now retired from the Ontario Ministry of Natural Resources, and Bill Pearce, New York Department of Environmental Conservation, co-chaired the steering committee which guided development of the Joint Strategic Plan for Management of Great Lakes Fisheries. Murray Johnson, Fisheries and Oceans Canada, is vice-chairman of the GLFC Habitat Advisory Board chaired by Bill Pearce. John Cooley, Fisheries and Oceans Canada, and Bob Pacific, U.S. Fish and Wildlife Service, co-chaired the team which developed the Guidelines for Fish Habitat Management and Planning in the Great Lakes. Ruth Koerber worked closely and diligently with the author through several manuscript versions.

REFERENCES

Baldwin, N.S., R.W. Saalfeld, M.A. Ross and H.J. Buettner 1979. Commercial fish production in the Great Lakes 1897-1977. Great Lakes Fish. Comm. Tech. Rep. 3: 187pp.

Berst, A.J. and G.R. Spangler 1973. Lake Huron - the ecology of the fish community and man's effects on it. Great Lakes Fish. Comm. Tech. Rep. 21:41p.

Canada Department of Fisheries and Oceans, Illinois Department of Conservation, Indiana Department of Natural Resources, National Marine Fisheries Service, New York State Department of Environmental Conservation, Ohio Department of Natural Resources, Ontario Ministry of Natural Resources, Pennsylvania Fish Commission, U.S. Fish and Wildlife Service, Wisconsin Department of Natural Resources 1980. A joint strategic plan for management of Great Lakes fisheries. Great Lakes Fish. Comm. 23p.

Christie, W.J. 1973. A review of the changes in the fish species composition of Lake Ontario. Great Lakes Fish. Comm. Tech. Rep. 23:65p.

Christie, W.J., M. Becker, J.W. Cowden and J.R. Vallentyne 1986. Managing the Great Lakes Basin as a home. J. Great Lakes Res. 12(1):2-17.

Eshenroder, R.L., T.P. Poe and C.H. Olver (ed.) 1984. Strategies for rehabilitation of lake trout in the Great Lakes: Proceedings of a conference on lake trout research. Great Lakes Fish. Comm. Tech. Rep. 40:63p.

Fetterolf, C.M. Jr. 1985. Lake trout futures in the Great Lakes. p. 1588-1593. In Richardson, F. and R.H. Hamre (ed.). Wild Trout III: Proceedings of the Symposium (Sept 24-25, 1984), Yellowstone National Park). The Federation of Fly Fishers and Trout Unlimited. 192p.

GLFC (Great Lakes Fishery Commission) 1983. Great Lakes Fisheries - Convention between the United States of America and Canada. Great Lakes Fish. Comm.

GLFC (Great Lakes Fishery Commission) 1986. Guidelines for fish habitat management and planning in the Great Lakes. In review. Great Lakes Fish. Comm.

Hartman, W.L. 1973. Effects of exploitation, environmental changes, and new species on the fish habitats and resources of Lake Erie. Great Lakes Fish. Comm. Tech. Rep. 22:43p.

IJC (International Joint Commission) 1965. Rules of procedures and text of treaty. International Joint Commission, Ottawa, Canada - Washington D.C.

IJC (International Joint Commission) 1978. Great Lakes Water Quality Agreement of 1978. International Joint Commission, Ottawa, Canada - Washington, D.C. - Windsor, Ontario.

Lamsa, A.K., C.M. Rovainen, D. Kolenosky and L.H. Hanson 1980. Sea lamprey (Petromyzon marinus) control - where to from here? Report of the Sea Lamprey International Symposium (SLIS) Control Theory Task Force. Can. J. Fish. Aquat. Sci. 37:2175-2192.

Lawrie, A.H. and J.F. Rahrer 1973. Lake Superior - a case history of the lake and its fisheries. Great Lakes Fish. Comm. Techn. Rep. 19:69p.

Pearce, W.A., R.A. Braem, S.M. Dustin and J.J. Tibbles 1980. Sea lamprey (Petromyzon marinus) in the Lower Great Lakes. Can. J. Fish. Aquat. Sci 37: 1802-1910.

Smith, B.R. and J.J. Tibbles 1980. Sea lamprey (Petromyzon marinus) in Lakes Huron, Michigan and Superior: history of invasion and control, 1936-78. Can. J. Fish. Aquat. Sci. 37:1780-1801.

Smith, S.H. 1972. Factors of ecological succession in oligotrophic fish communities of the Laurentian Great Lakes. J. Fish. Res. Board Can. 29;717-730.

Talhelm, D.R. 1986a. Economics of Great Lakes fisheries: A 1985 assessment. Executive Summary. Minutes of the Annual Meeting of the Great Lakes Fishery Commission, May 1986. In press. Full version accepted for publication by Great lakes Fishery Commission.

Talhelm, D.R. 1986b. The international Great Lakes sport fishery of 1980. Great Lakes Fish. Comm. Spec. Pub. 86-2. In press.

Wells L. and A.L. McLain 1973. Lake Michigan - Man's effects on native fish stocks and other biota. Great Lakes Fish. Comm. Tech. Rep. 20:55p.

CHAPTER 6

LAKE ORTA (N. ITALY): RECOVERY AFTER THE ADOPTION OF RESTORATION PLANS

C. Bonacina[1], G. Bonomi[1,2], L. Barbanti[1], R. Mosello[1], D. Ruggiu[1] and G. Tartari[3]

[1] CNR-Istituto Italiano di Idrobiologia, Largo Tonolli 50;I 18048, Pallanza NO (Italy)

[2] Universita di Bologna, Dipartimento di Biologia Evoluzionistica e Sperimentale,Cattedra di Idrobiologia e Pesicoltura, Via S. Giacomo 9, I 40126 Bologna BO (Italy)

[3] CNR-Istituto Ricerca Sulle Acque, Via delle Cascine Occhiate, I 20047 Brugherio MI (Italy)

ABSTRACT

Heavy industrial pollution (copper and ammonium sulphate) from a rayon factory was responsible for the disappearance of almost all forms of life in the lake since the late twenties. The in-lake $N\text{-}NH_4$ oxidation produced a gradual $N\text{-}NO_3$ accumulation and a progressive strong decrease of pH down to values around 4. The copper concentration peaked at about 100 ug/L in the mid sixties, when additional sources of heavy metals (Cu, Zn, Cr) i.e. several new metal plating factories were set up on the lake western shores. Toxicity texts with rainbow trout indicated that copper at the prevailing very low pH was the major cause of absence of pelagial fish in the lake.

In 1976 the new Italian water pollution law obliged the rayon factory to set up a large, new treatment plant for the recovery of copper and ammonium sulphate; at the same time new plants for domestic and industrial effluents were planned and are now finished or under construction. This resulted in an immediate change in the lake water composition, particularly in the $N\text{-}NH_4$ concentration (now at about 2 mg $N\text{-}NH_4$/L) and in some signs of recovery in the biological community, e.g. blooms of *Brachionus urceolaris* and the settlement of a new population of *Tubifex tubifex* in the profundal zone.

A research program is now being conducted on the lake and its "paralimnion", in order to adequately survey the recovery process and to give useful suggestions for the adoption of possible direct measures. Liming of the South basin of the lake is being seriously taken into consideration and a proper scientific and technical program set up as a collaboration between the CNR-Istituto Italiano di Idrobiologia, Pallanza and the Regione Piemonte and the Provincia di Novara.

It is an honor to be here at this World Conference on Large Lakes and have the opportunity to show the case of Lake Orta, with particular reference to its recent striking recovery. We feel that the results obtained and the convergence of the chemical and biological situation of our lake with those lakes in the world, that similarly undergo acidification as a result of acid depositions, make the Lake Orta case one of very general interest.

INTRODUCTION

Lake Orta is the seventh largest Italian lake by volume (1.3×10^9 m^3 and depth (max: 143 m; mean: 71 m). It occupies the south-western part of the larger Lake Maggiore drainage basin. Its outlet, River Nigoglia (Figure 1) joins the River Strona, a tributary of the River Toce, which flows into Lake Maggiore.

Both Lake Orta and its drainage basin have a long, narrow shape extending in a N-S direction. The main sub-basins are located in the western part of the area, except for that of the River Pescone (18 km^2), and are formed by the contributing areas of the following streams (from North to South): rivers Bagnella, Acqualba, Pellino and Lagna (Figure 1).

Some characteristics of Lake Orta are given in Table 1. It is very important to note that the actual mean water residence time is 10.7 years. It has been calculated that after 10.7 years, 45 % of the original water is

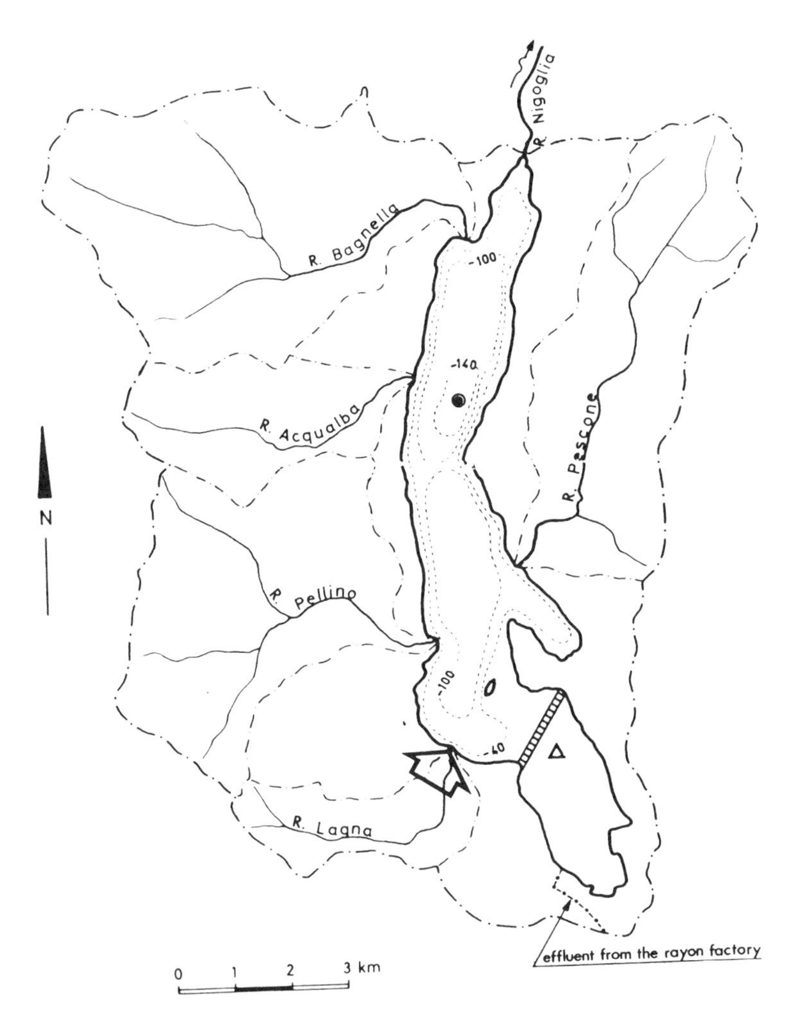

Figure 1. Sketch map of Lake Orta an its drainage basin. The circle indicates the normal station for chemistry and plankton sampling; the triangle, the station for profundal benthos; the barred strip indicates the transect: Imolo-Punta Casario, along which the drift crosses are released for water current studies; the large arrow indicates the location of the main complex of plating factories.

TABLE 1 SOME HYDROGRAPHICAL AND LIMNOLOGICAL FEATURES OF LAKE ORTA

Drainage Basin Area	116 km^2
Mean Altitude of Drainage Basin	650 m a.s.l.
Mean Lake Level Altitude	290 m a.s.l.
Lake Area	18.136 km^2
Lake Volume	$1286x10^6$ m^3
Mean Depth	70.9 m
Maximum Depth	143 m
Water Inflow (precipitation)	1901 mm/yr
Mean Outflow Discharge	4.81 m^3/s
Yearly Outflow Discharge	$151.688x10^6$ m^3
Theoretical Detention Time	8.478 yrs
Theoretical Renewal Rate	0.118 /yr
Actual Renewal Rate	0.093 /yr
Mean Actual Water Residence Time	10.7 years

still present in the lake (Bonacina and Bonomi 1974). The lake basin, formed by ice erosion on a pre-existing river valley, is mostly formed of gneiss, micashists and granites. Because of the geology of the catchment basin the water of Lake Orta was originally poorly buffered, with total alkalinity ranging from 0.3 - 0.4 meq/L (Monti 1929).

The very pure waters of Lake Orta once supported a rich biological community: about 150 algal species (Giaj-Levra 1925) and an abundant population of copepods and, above all, of cladocerans (Pavesi 1379, Monti 1929) which are the favorite food items for zooplanktophagous fishes. Fish were also abundant: trout (Salmo lacustris), shad (Alosa finta lacustris), bleak (Alburnus alburnus alborella), perch (Perca fluviatilis), pike (Exos lucius) and eel (Anguilla anguilla) were the major catch. After 1901 Coregonus wartmanni (white fish) was successfully introduced in the lake and became one of the most important fishes from the economic point of view. In 1914-1916 Salvelinus alpinus (char) was also introduced (De Agostini 1925). In 1911 a Fisherman's Cooperative Association was founded with the aim, among others, of taking care of fish reproduction and restocking. The yield of fish in Lake Orta was very high: in 1927 an issue of the Proceedings of the Italian Ministry of Agriculture, Industry and Commerce reports the total yearly catch to be 65.4 tons, given as a normal production for the lake; that is a production of 36.2 kg/ha, almost twice that of Lake Maggiore (20.5 kg/ha; Baldi 1979), at that time.

Bearing in mind the lake volume of Lake Orta (Table 1) and its value as a holiday resort and a centre for water sports, it is readily concluded that its restoration and protection are high desirable.

THE CASE HISTORY

In November, 1926, a rayon factory (Gemberg, Figure 2b) was set up at the southern end of the lake (the outlet, River Nigoglia, leaves the lake at its northern end, Figure 1). The factory needed (and still does) a large quantity of pure soft water for washing and cooling purposes. The factory discharge is "enriched" with copper sulphate and ammonium sulphate, at a pH of about 10 and temperature of 18-22°C. Some iron salts were also discharged, due to a process of partial copper recovery. The mean daily discharge of the factory effluent was, and still is, 12,000 m^3. This represents about 3% of the lake mean discharge.

At the end of 1927, water samples from the open lake were completely devoid of phyto- and zooplankton (Bachmann, in litteris). Two years later, practically all forms of life had disappeared, and the lake water was classified as sterile (Monti 1930). According to Monti, this extraordinary chain of events was attributable to the algicidal action of copper, which rapidly destroyed almost all the phytoplankton and consequently zooplankton and fish. The factory which caused this ecological disaster tried in many ways to deny its responsibility. The inhabitants were divided between fishermen deprived of their livelihood and the people who had found work in the factory. Eventually, all fishermen were employed by the factory and the controversy stopped. However, the problem of the lake remained.

The factory effluent contained, in varying quantities through the years, ammonium and copper sulphate (Table 2), plus other substances such as sodium carbonate and calcium oxide which are not relevant in this context. The decrease in the total inorganic nitrogen loading during the period 1975-1979 was due to a decline in the rayon market. As shown, in the late fifties, the total amount of copper reaching the lake increased greatly. This is attributed to the many small bathroom accessory plating factories which opened on the south-western shore near the mouth of the River Lagna (Figure 1 and 2d). They discharged heavy metals (Cu, Zn, Cr, Ni), anionic detergents and cyanide into the lake. The lowering of copper loadings in the period 1959 to 1967 L (Table 2) was due to the installation of a copper recovery plant in 1958.

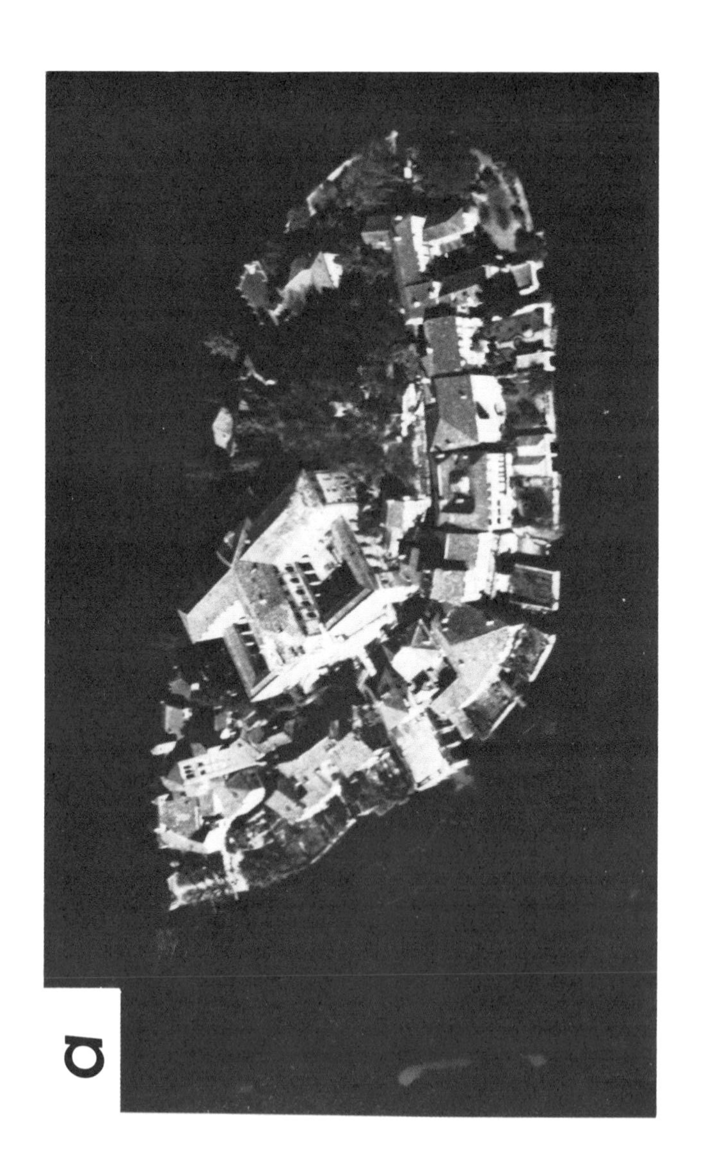

Figure 2a. Aerial view of Isola S. Giulio (Central Basin).

Figure 2b. Aerial view of the Bemberg factory area located immediately to the south of Lake Orta.

Figure 2c. Aerial view of the Bemberg factory area with the factory lake water reservoirs used for rayon process water.

TABLE 2 ANNUAL TOTAL INORGANIC NITROGEN (TN_{in}) AND TOTAL COPPER LOADING FROM THE RAYON FACTORY TO THE LAKE (1927-1979)

PERIOD	TOTAL INORGANIC NITROGEN (TN_{in}) tons/yr	COPPER (tons/yr)
1927-1946	up to 1,000	up to 40
1947-1958	1,000-2,000	40-80
1959-1967	2,000-3,000	4- 5
1968-1974	about 3,000[1]	about 4
1975-1979	about 2,000	about 3

[1] max.3,350 in 1970

For copper, the mean annual loading from these factories of was calculated to be 13 tonnes (Bonacina and Ruggiu 1973). As far as heavy metals are concerned, the available data are rather scanty. The mean concentrations at Station A (Figure 1) are: approximately 40 ug total Cu/L, 60 ug total Zn/L and 5 ug total Cr/L. About 85 ug Al/L were also found (the Al is likely leached from the lake sediments). The situation is made worse because of metal accumulation in the sediments (Corbella et al. 1958, Barbanti et al. 1972, Bonomi and Ruggiu 1974). Recently Provini and Gaggino (1985) calculated that the metal content of the lake sediments, particularly chromium, constitutes a "considerable" to "very high" ecological risk.

As a result of this heavy loading, nitrates (derived from in-lake ammonia oxidation) started to accumulate (Baldi 1949) and pH to decrease to about 4 while the buffer capacity of the lake tended to zero (Vollenweider 1963). Year by year the oxygen decreased due to nitrification processes. Finally, the large quantities of ammonia were no longer completely oxidized and started to accumulate in the lake waters, while the oxygen depletion in the hypolimnion increased (Bonacina 1970).

At the beginning of the seventies, the situation at the circulation period was as follows: mean concentration of ammonia nitrogen: 4.8 mg/L; nitrate nitrogen: 5.5 mg/L; copper: 50 ug/L; pH: 4.5; oxygen saturation: 60% (Barbanti et al. 1972). During the summer, due to its relatively high temperature, the factory effluent tended to stratify within the epilimnion. As a consequence, ammonia concentration increased to as high as 8 mg N/L in the first ten meters (some 1 mg/L of which is free ammonia), while pH ranged from 7-9 (Bonacina 1970, Barbanti et al. 1972, Bonacina and Bonomi 1974). In addition to this vertical stratification, a horizontal south to north gradient was present from spring to autumn. This was due to the relative positions of the polluted inlet and outlet. Indeed, the epilimnic summer concentration of ammonia nitrogen was higher near the southern end of the lake (from south to north: 8.5 and 6.6 mg N/L) whereas epilimnic summer values of pH varied in the same direction from 8.9 to 8.4 (Bonacina and Bonomi 1974).

The biological situation had changed slightly since 1927. In fact, sudden algal and rotifers blooms started no later than August 1929 (Bachmaniella planctonica Chodat, nov. gen., nov. sp.; Baldi 1949). Afterwards there was a burst of Pedalia mira (= Hexarthra fennica) in September 1946 (Baldi 1949) and later on, in July 1971 (Barbanti et al. 1972) successively, there were blooms of Scenedesmus armatus (Nov.-Dec. 1960; Vollenweider 1963) Ankistrodesmus convolutus (Jul. 1969; Bonacina 1970), Brachionus urceolaris (Oct. 1983 and Jan. 1984; Bonacina, unpublished data). In the seventies, there were very few planktonic species inhabiting Lake Orta. There were two algal species: Oscillatoria limnetica (blue green) and Coccomyxa minor (green), plus some diatom species with very low density, one copepod species (Cyclops abyssorum) and some rotifer species (Brachionusurceolaris, B. calyciflorus, Keratella cochlearis). In the profundal of the lake there were no macrobenthonic species but a rich population of benthic ciliates flourished (Ruggiu 1969).

There were no fishes at all. In 1975 short-term field survival tests showed ionic copper to be the major acute toxic factor for adult specimen of the rainbow trout (Salmo gairdneri) (Calamari and Marchetti 1975). However, there are also chronic effects which must be borne in mind: investigations on Norwegian lakes have demonstrated a strong correlation between lowering of pH and progressive lack of fish fauna in low alkalinity waters (Muniz et al. 1984). Furthermore, it must be remembered that other metals are permanently present in the lake waters. Their toxicity to fish fauna and zooplankton is well known (Haux and Larsson 1984, Intersool and Winner 1982, Winner 1976 and 1984).

RECOVERY ACTION

For the successful restoration of Lake Orta there were three main problems which had to be solved:

1. stopping the continuous input of ammonia nitrogen,

2. treating the heavy metal wastes from the plating factories (Figure 2d), and

3. preventing the sewage from 35,000 inhabitants (tourists included) from reaching the lake.

The first problem has ceased to exist. The rayon factory, compelled by the Italian law on water pollution (approved in 1976 by the Italian Parliament) constructed a big treatment plant for the recovery of ammonia and copper salts. Operation started in late 1980. Consequently, the mean annual loading (Cu and TN_{in}) from the factory effluent since 1980 has been about 0.3 tons/yr and 30 tons/yr respectively.

The second problem is still awaiting solution. Although a treatment plant has been constructed (Figures 2f and 3) it has not yet started to work effectively. This is because it is an activated sludge plant, which receives both domestic sewage and effluents from the plating industries, which unfortunately are still characterized by peak metal concentrations that inhibit the proper operation of the plant. Table 3 reports the estimated heavy metal loadings from the plating industries.

The third problem will be solved shortly when both the new Omegna treatment plant (Figure 2e and 3) and the villages to be served are connected with the pipeline network.

SIGNS OF RECOVERY

Some significant effects have been produced on the lake waters by the recovery of ammonium sulphate from the factory effluent. Firstly, there has been a decrease in the total inorganic nitrogen (TN_{in}) concentration, in particular that of ammonia nitrogen. It has dropped from its early 1980 level of about 5 mg/L to about 2 mg/L. Secondly, there has been a marked decrease of

Figure 2d. Aerial view of area where most of the plating factories are located.

Figure 2e. Wastewater treatment plant for the town of Omegna (see P1 on Figure 3) which shall receive domestic (mainly) and industrial wastes.

Figure 2f. Wastewater treatment plant on the Lagna stream, receiving sewage from the nearest villages and industrial wastewater from the plating factories (see P2 on Figure 3).

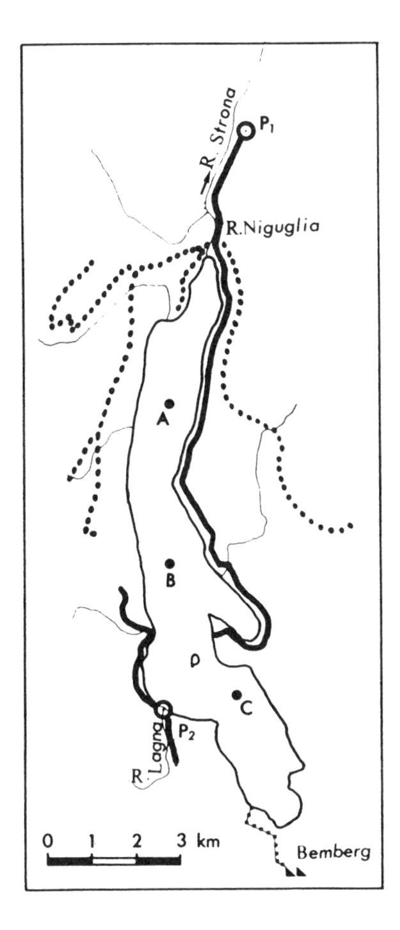

Figure 3. Lake Orta and Omegna (P1) and Lagna (P2) wastewater treatment locations and collection network. Dotted lines indicate future additional collection sewers.

summer-autumn epilimnic N-NH_4 concentration and pH values. The third effect is the abatement of the horizontal north to south N-NH_4 and pH lake-long gradient (Bonacina and Bonomi 1974).

The TN_{in} load received by the lake today is less than 1% of the previous peak leads, resulting in the development of new vertical and horizontal stratifications. This causes lake water pH now to be in the 4-6 range year round and from surface to bottom. An inversion in the previous horizontal gradients seems

TABLE 3 PRESENT COPPER AND CHROMIUM LOADINGS TO THE LAKE:
A: conveyed to the Lagna treatment plant
B: directly disposed of into the lake

METAL	YEAR	A (tons/yr)	B (tons/yr)
Copper	1983	about 4	about 1
	1984	0.7	0.5
	1985	0.7	0.5
Chromium	1983	9.1	5.5
	1984	9.0	5.5
	1985	2.4	5.5

to be developing along the main axis. The nitrification of the residual mass of ammonium ion in the lake and the lack of the bicarbonate buffer system are responsible for the still very low pH values. In spite of the apparently rather low nitrification rate, calculated by some means of hypolimnic mass-balances (Bonacina and Ruggiu 1973, Gerletti and Provini 1978) and the low nitrifier counts (Gerletti and Provini 1978) the lake as a whole is showing at the moment the effects of efficient and marked nitrification. In this context it must be borne in mind that during the sixties the loading into the lake from the rayon factory was about 3,000 tons $N\text{-}NH_4$/yr. This resulted in a sharp increase in the $N\text{-}NH_4$ concentration (from 1 to 4 mg/L in 6 years) and a parallel decrease in the concentration of the lake water nitrate (Bonacina 1970). The limits of the nitrification potential of the lake system were apparently thus shown. The authors are of the opinion that the current $N\text{-}NH_4$ concentration and the comparatively insignificant $N\text{-}NH_4$ loading from the rayon factory will maintain a considerable capacity for nitrification and will result in a rapid decline of $N\text{-}NH_4$ in the lake waters, but in a slow decrease in nitrates. Consequently, the pH will tend to stay very low very several years.

Some immediate responses of the biota to the rapid chemical changes have already occurred. There are occasional blooms of some sporadic planktonic populations, more now than in the past, e.g. *Scenedesmus* sp., *Brachionus urceolaris*. Some typically littoral cladocerans (*Syda cristallina*, *Chydorus sphaericus*) may now be found in open waters. Recovery by the profundal zones of all three lake basins (Figure 1) through colonization by a macrobenthonic population of the tubificid *Tubifex tubifex* is becoming evident.

A possible recovery of the fish community has not yet been documented. It is anticipated that this community will probably suffer even further setbacks as a result of the extreme acidity, the somewhat high heavy metal concentrations and the current new epilimnic situation (low pH for the whole year). Recently the authors were unable to detect any fish mark with their echosound recorder when using this instrument in many areas of the lake, including the main inflowing stream inlets.

CURRENT RESEARCH

Hydrography of Lake Orta

The paramount morphological characteristic of Lake Orta is its subdivision into three different basins with a progressively increasing depth from south to north. From the hydrographic point of view it is different from the other deep Italian sudalpine lakes in that its effluent, River Nigoglia, outflows from the northern end of the lacustrine valley. The area for dumping copper and ammonia waste into the lake, responsible for the past large scale chemical pollution, is located at the opposite end (Figure 1).

In this geo-hydrographic context, the CNR-Istituto Italiano di Idrobiologia began a study on water movements in lake Orta, especially in its southern basin, with three different aims:

1. to investigate the possibility that the pollutants trapped in the lacustrine sediments close to the rayon factory discharge might be released and transported by water movements into other parts of the lake;

2. to analyze the diffusion of the present wastewater from the industrial factory into the lake;

3. to verify the possibility of using suitable techniques in order to speed up the recovery rate of the lake, for instance, liming of its southern basin, or part of it.

The first approach to the water movements of Lake Orta was carried out using current crosses released at five depths (0.5, 10, 20 and 30 m) along a fixed transect:

Imolo-Punta Casario (maximum depth 34 m; Figure 1). During each field experiment (five in 1985 and two so far, in 1986) the crosses were topographically positioned at intervals of approximately two hours. It was recognized that wind is the most important acting force, especially for water in the upper levels. In these strata the water velocity was, in some cases, more than 10 cm/s. The deeper the levels investigated, the less swift the water movements, so near the bottom the maximum speed was 3-4 cm/s. The roles of thermal stratification of the morphology of the lacustrine valley in determining the energetic and directional components of water currents were also recognized. Their maximum velocity was recorded during the spring 1986 when at the beginning of stratification, the work of the wind was distributed exclusively on the first strata, the only ones involved in the circulation of the lake. The general current pattern is therefore linked with the present anemological situation, but it is also affected by the former winds.

Generally speaking, the counterbalancing movements occur at the same depth, but along opposite shores. The most common hydrological situation in Lake Orta indicates an anti-clockwise circulation with currents flowing south or south-east along the western shore, counterbalanced by a northward movement along the eastern shore. However, the opposite situation has also been recorded, i.e. a clockwise circulation. Counter-balancing at different depths is also possible, generally with the more superficial strata flowing north and the deepest in the opposite direction.

The next step in the research program will deal with the process of diffusion of the industrial effluent into the lake. The effluent discharge rate is about 12,000 m^3/d with a temperature of 25-30^oC. If the effluent does not contain some physical or chemical characteristic which is sufficiently conservative to permit a suitable measurement of turbulent diffusion, it will be necessary to use a tracer, i.e. fluroescent dye.

Chemistry of Lake Orta

Though the atmospheric deposition in the watershed of Lake Orta is decidedly acidic (Table 4) and constitutes a hydrogen ion input on the lake surface (18.12 km^2), the water of the tributaries (draining altogether 97.6 km^2) are buffered (Table 4). The mean total alkalinity

TABLE 4 pH, CONDUCTIVITY AND MEAN CONCENTRATIONS OF ATMOSPHERIC DEPOSITIONS (A) AND INFLOWING WATERS (B); mean values for the years 1984-1985, compared with the lake chemistry at the overturn (C:1984; D:1985)

PARAMETER		A	B	C	D
pH	-	7.9	7.5	4.4	3.9
Conductivity	uS/cm	29	90	124	153
H^+	ueq/L	39	0	44	126
NH_4^+	ueq/l	58	25	284	164
Ca^{++}	ueq/L	29	303	315	324
Mg^{++}	ueq/L	6	117	123	123
HCO_3^-	ueq/L	0	308	0	0
SO_4^{--}	ueq/L	83	223	641	645
NO_3^-	ueq/L	48	93	319	321
SUM cations[1]	ueq/L	146	693	1031	1013
SUM anions[2]	ueq/L	146	704	1036	1032

[1] includes Na^+, K^+ [2] includes Cl^-

is about 300 meq L^{-1} and the pH generally above 7.0. Even during heavy rains, which are quite frequent in this area, no episode of acid input from the tributaries has been detected (Mosello, unpublished data). So, it is evident that the strong acidification of the lake water is due to in-lake processes.

Many in-lake reactions, mostly biologically mediated, produce or consume hydrogen ions (e.g. Stumm and Morgan 1981, Brewer and Goldman 1976, Schindler et al. 1985). In the case of Lake Orta, ammonia oxidation has been recognized since 1962 as a source of acidity (Vollenweider 1963 and 1965). This has also been recently confirmed on the basis of a chemical budget (Mosello et al. in press). The lake water showed a pH of 4.0-4.5 during the years 1984-1985, with minima of 3.8-3.9 occurring at the end of the stratification of 1984. Table 4 shows the mean pH and ionic concentration at the lake overturn. The surface water shows lower acidity and pH may reach values as high as 6.0 during summer. This is due to the influence of tributaries and biological processes of phytoplankton.

It is to be stressed that the present chemical condition of the lake is very different from its original state, before the beginning of ammonia pollution, as documented from earlier studies (Monti 1930, Baldi 1949) as mentioned in the introduction of this paper. The same conclusion may be obtained by analogy with the chemistry of a lake located in the same region, Lake Mergozzo, hydrogeochemically very close to Lake Orta.

The relation between ammonia oxidation and pH decrease is shown in Figures 4a and 4b, respectively.

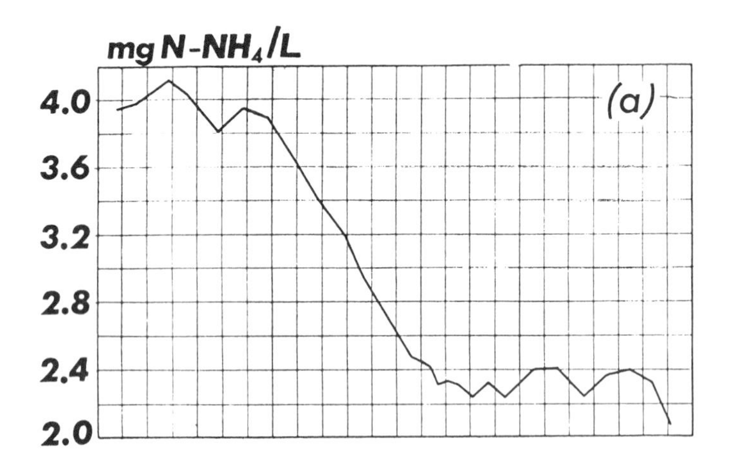

Figure 4a. Ammonium nitrogen in Lake Orta (1984-1985). (Values are means calculated for the 0-143m water column).

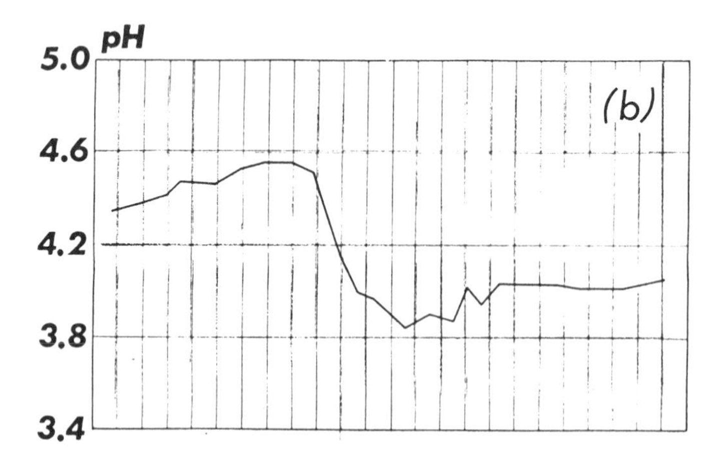

Figure 4b. pH values in Lake Orta (1984-1985). (Values are means calculated for the 0-143m water column).

The ammonia nitrogen decreased substantially from 1985, practically halving the values in the lake. This decrease is of great importance when it is also compared with the long-term variation of ammonia concentration (Figure 4). The more relevant pH decrease was from September 1984 to January 1985. Also, in this case the values are among the lowest measured in the lake. The oxygen concentration shows a decrease as well. This is in agreement with the decrease in ammonia concentration. During 1985 ammonia concentration, though again very high (2.0 - 2.4 mg N/L), showed a decrease much lower than the previous year (Figure 4). The variation of pH and oxygen concentrations during 1985 are consistent with that of ammonia, showing in the case of pH a moderate increase in comparison with the overturn value, and in the case of oxygen (Figure 4c), a lower decrease in 1985 than in 1984.

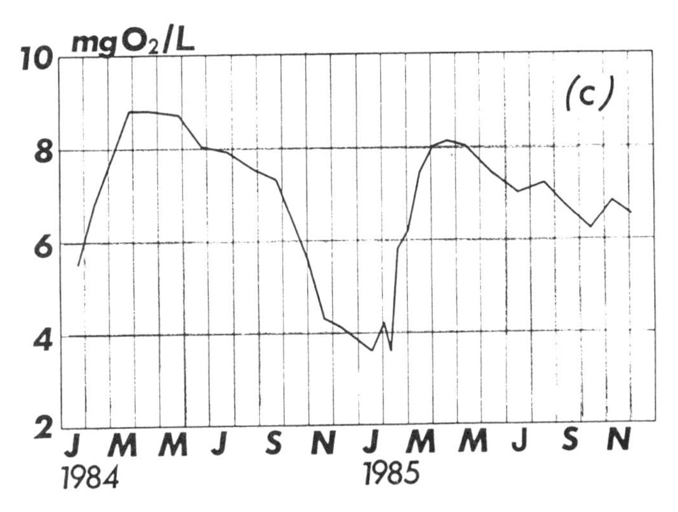

Figure 4c. Dissolved Oxygen values in Lake Orta (1984 -1985). (Values are means calculated for the 0-143m water column).

The slight increase in pH during 1985 is related to the alkalinity input from the watershed, as calculated from the chemical budget (Mosello unpublished). The lower ammonia oxidation rate during 1985 is probably due to the inhibitory effect of pH, which in the hypolimnion was around 4.0 year round.

The acidity of the lake waters permits a relatively high concentrations of heavy metals in solution which are derived from the industrial activities in the watershed. At the 1985 overturn, the concentrations were 36 ug Cu/L, 80 ug Al/L and 55 ug Zn/L.

Phytoplankton in Lake Orta

It would be interesting, and rewarding, to give a full account of algal assemblages in Lake Orta from the time of the great pollution of 1926 to the present. It is a history of community collapse with great reduction in number of species, mass mortality of diatoms, recurrent ephemeral bursts of various species disconnected with the pristine communities, and no clear and established seasonal successions. As these events are beyond the scope of this discussion, reference is made to the papers of Vollenweider (1963) and Bonacina (1970), where the lake conditions of the past were extensively treated. Suffice it to say that by 1970 only two species, the blue-green *Oscillatoria limnetica* (then described as *Lyngbya limnetica*) and the green *Coccomyxa minor* comprised the largest part of the phytoplankton throughout the year. By 1981, just after the wastewater treatment plant came into operation, some additional species were of importance (*Scenedesmus armatus*, *Microcystis aeruginosa*, *Achnantes minutissima*). A seasonal succession was evident.

Detailed observation on phytoplankton are currently under way, but still remain to be elaborated. However, after five years of treatment plant activity, things seem to be changing considerably and quickly. *C. minor* has further increased its importance, but the blue-greens have dwindled almost to nothing and, more important, a lot of small flagellate species belonging to several classes have appeared, together with some minute diatoms. It is tempting to conclude that the rapidly changing chemical conditions are flavoring the occurrence of opportunistic, pioneer species and to predict major changes in community structure in the future.

The authors believe that the peculiarities of Lake Orta offer a unique opportunity of gaining information on the ecology of phytoplankton, as the situation closely resembles a large-scale experiment. For this reason, the chemical and physical parameters are being monitored in the water column, along the plankton sampling and primary production measurement with the refinement of the oxygen technique. On the other hand, bioassay experiments are under way to test specific hypotheses on the effect of changing pH and nutrient supply, and of the presence of heavy metals on the community metabolism and composition.

The authors also realize that practical suggestions for lake management will be a necessary outcome of their activity and their studies are largely planned with this end in mind.

Cycle and Toxicity of Copper in Lake Orta

Since the middle of the seventies, the Laboratory of Applied Hydrobiology of the Water Research Institute, in Brugherio (Milano), has investigated at different times, specific topics concerning Lake Orta: the acute toxicity of copper and ammonia (Calamari and Marchetti 1975), the nitrification processes (Gerletti and Provini 1978) and copper content in sediments (Provini and Gaggino 1985). At present, the copper cycle and its transport mechanisms are being investigated. The copper partitioning in dissolved and particulate phases has been investigated in order to assess the metal cycle in the lake, whose hydrochemical conditions have perceptibly changed during latest years (Mosello et al. in press). The soluble forms of copper, in the dissolved phase, are characterized after filtration on polycarbonate membrane (0.4 um). The fractions of the particulate phase are studied in the suspended particulate matter and the fresh sedimenting material which is caught monthly in two traps (30 and 140 m depth). The former is separated by centrifuging water samples (20 L) with a continuous flow system, then both are wet digested with nitric and hydrofluoric acids by heating in vessels (PTFE-lined).

The results show that copper, at the present conditions of pH 4.5 year round in the whole water column and of dissolved organic carbon (DOC) of 1 mg/L, is mostly (98%) as the ionic species in the dissolved phase. This has been confirmed by electrochemical analyses. The sedimentary fluxes of particulate matter range from 0.3 to 6.3 $g/m^2.d$ in euphotic zone and from 0.5 to 10.1 $g/m^2.d$ at the bottom, with minimum values during a large part of the year.

These results, together with those of further investigations that simulate changes of the environmental conditions , may be useful to elucidate the transport mechanisms copper and its behaviour in the lake. The results may also be used in a mathematical model.

Zooplankton in Lake Orta

After the catastrophic destruction of 1927 (Monti 1930) zooplankton of Lake Orta never returned to the previous normal condition of a balanced community. Like phytoplankton, zooplankton also underwent episodes of instantaneous blooms of various species (see also the section "case history"), but the bulk community was made up by a cyclopoid copepod, Cyclops abyssorum, still present with very few individuals (Vollenweider 1963, Banocina 1970). Today the zooplankton of Lake Orta is intensively studied. The main sampling point is Station A (Figure 3) where bimonthly samples are taken by means of a Clarke and Bumpus plankton sampler, and then examined and counted in order to update species composition and abundances. For these purposes, nets with differentiated meshes are used: in this way adult (bigger) and young (smaller) copepod and cladoceran forms as well as the small rotifers can be caught. In order to obtain specimen of rare species, samples are taken occasionally at various stations with much larger, although with the same mesh size, nets.

In 1985-1986 the most important species, from the numerical point of view, has been Brachionus urceolaris (Mueller) (Rotifera, Brachionidae) whose population can reach a density of 200,000 m^{-3} (Bonacina Jan. 1984 unpublished).

The Profundal Macrobenthos

In 1983, during one of the periodical surveys performed on the lake, a population of Tubifex tubifex Mueller (Oligochaeta; Tubificidae) was discovered in the profundal of Lake Orta. These worms are distributed all over the profundal of the three basins (from 20 to 143 m depth). In the bottom layers from the eu-littoral to about -20 m several (10) stonefly species are present, together with 2 chironomids: a Chironomus sp. and a Procladius sp..

During 1984, at a fixed station in the south basin (Station B, Figure 3) 12 replicate Ekman samples were taken at monthly intervals. The sorted material was divided into 5 compartments (eggs, embryos, young, mature and ovigerous individuals). The population displays a marked clumped distribution, typical of a new settler. The overall abundance values may be as high as 37,000 m^{-2}.

The clutch size (eggs/cocoon) of this population is very high. Marked mid-summer mortality has been noted. An annual numerical balance for the 5 biological compartments shows that the "cost" of one ovigerous individual, in terms of mortality of the previous compartments, does not sustantially differ from that estimated for other unpolluted lakes (Bonomi and Adreani 1978, Andreani et al. 1981, 1984).

EXPEDIENCY OF TAKING DIRECT MEASURES

The strong mineral acidity of the lake waters and the relatively high concentration of some heavy metals are the true bottleneck of the Lake Orta ecosystem at the moment. Ironically enough, the present pH situation is even worse because of the lake alkalinity from the rayon factory which, up to some years ago, had some beneficial effects on the epilimnic waters (e.g. increasing pH and precipitating Cu).

Therefore liming the lake appears to be a possible direct measure, naturally taking into account the feasibility and cost of such a scheme. This measure was already started some 25 years ago, but its cost (large lake volume) and a need for a non-stop intervention (continuous past heavy NH_4 loading, producing a continuous long term acidification) left the idea at a theoretical stage. Furthermore, in the past situation, when a wastewater treatment plant for domestic sewage was not even contemplated, an inorganic C addition (liming) and the P loading (untreated sewage) might have transformed the lake into a very productive system. The present situation looks drastically different. The actual acidity of the lake water and the potential proton production resulting from the nitrification of the residual ("historical") 2 mg $N\text{-}NH_4/L$ may easily be calculated and the appropriate amount of carbonate required can then be computed and added to the lake waters. Chemical addition may be performed in several stages but need not be repeated over a very long period of time. Being aware of technical problems and of the high cost of liming the whole lake volume Bonacina and Bonomi (1984 and 1985) suggested limiting the liming to the southern basin, which has a much smaller volume.

A couple of years ago this proposed direct measure was still regarded by the local administrators as a "scientific curiosity", as the lake was in fact recovering anyway and at a very fast rate. Very recently (spring 1986) the local health authorities realized that, due to the progressive pH decrease, the superficial waters were well below the pH range (6-9) allowed for public bathing. Consequently bathing was prohibited and liming became a direct measure to be seriously considered by the authorities themselves. At the moment the Regione Piemonto (one of the twenty administrative units in Italy) is willing to sponsor the project, possibly delegating the Province of Novara (local administration subdivision of Piemonte) to prepare the program. Both the scientific and technical aspects are to be set up in strict collaboration with the CNR-Istituto Italiano di Idrobiologia, Pallanza.

The scientific and technical program is being prepared. The technical aspects should be facilitated by contact with institutions or environmental agencies that have carried out and continue to carry out liming of acidified water on a large scale (e.g. Scandinavian countries). The local (south end) liming should produce, in a reasonably short time, the following striking effects:

1. neutralization of part of the lake and reconstitution of a "normal" alkaline reserve in this basin,

2. precipitation of heavy metals with a consequent strong decrease in toxicity for the present lake biota and for potential re-settlers,

3. enhancement of the nitrification of the residual ammonia (nitrification is now probably pH and inorganic carbon limited), and

4. reconstitution of a much richer animal community, including zooplankton and fish. Fish re-stocking will have to be strictly controlled by the Fishing Authorities (Province of Novara).

A continuous mild liming of the south basin - a future possibility - should also be effective in enhancing the nitrification of the $N\text{-}NH_4$ load from the rayon factory. Indeed, although in the present chemical situation the factory loading (30-40 tons $N\text{-}NH_4$/year) is negligible in comparison to the total ammonia mass in the lake (approximately 2,600 tons $N\text{-}NH_4$). In a future situation, when the residual $N\text{-}NH_4$ is totally oxidized to nitrate, the factory $N\text{-}NH_4$ loading is likely to have some local effects in the south basin.

The turbulent diffusion (drift, internal seiches and convective currents) should help in gradually bettering the chemical situation in the central and north basins of Lake Orta (Figure 1).

CONCLUDING REMARKS

Undoubtedly the case of Lake Orta is a unique example of how uncontrolled industrial pollution may throw a large deep lake into acute stress. Its dramatic environmental deterioration, both chemical and biological, was immediately detected by observers of the lake and three generations of limnologists, almost all of them from the CNR-Istituto Italiano di Idrobiologia in Pallanza. They have worked with the aim of not only surveying the situation of the lake but, more importantly, of pushing towards its restoration. We have been fortunate enough to be present at the start and follow its amelioration. The scenario demonstrates how, even in heavily polluted large lakes, the implementation of adequate treatment and recovery plants leads to an immediate improvement in the environmental situation.

A last stage in the "resurrection" of Lake Orta could be the adoption of some direct measures, e.g. liming of at least part of its waters. We are eager to participate in this enterprise and hope to be able to follow this final acceleration of its recovery.

ACKNOWLEDGEMENTS

We are deeply indebted to the Governor of the State of Michigan for inviting G. Bonomi to attend the World Conference on large Lakes-Focus on Toxics, held at Mackinac Island, Michigan. We also acknowledge the role of Dr. H.F. Anderson (FAO, Rome) and of Prof. H. Regier (University of Toronto) in suggesting the invitation.

Thanks are due to the Italian National Research Council and to the province of Novara for providing partial funding of the research programs that are presently being conducted on the lake. We are grateful to Ing. A. Lanza and Dr. G.L. Triveri, the Bermberg Company, for providing data on the factory and easy access to its treatment plant; Arch. L. Rivetti and Dr. Lacqua, respectively President and Director of the Lagna and

Omegna treatment plants are also thanked for their kind collaboration. Our colleagues and technicians at the CNR-Istituto Italiano di Idrobiologia, Pallanza were very co-operative at all stages of the work.

The English text was kindly revised and partly re-written by Ms. Sandra Spence.

REFERENCES

Adreani, L., Bonacina, C. and Bonomi, G. 1981. Production and population dynamics in profundal lacustrine Oligochaeta. Verh. Internat. Verein. Limnol. 21:967-974.

Adreani, L., Bonacina, C. and Bonomi, G. and Monti, C. 1984. Cohort cultures of *Psammoryctides barbatus* (Grube) and *Spirosperma ferox* Eisen: a tool for a better understanding of demographic strategies in tubificids. Hydrobiologia 115:113-119.

Baldi, E. 1949. Il Lago d'Orta, suo declino biologico e condizioni attuali. Mem. Ist. Ital. Idrobio. 5;145-188.

Barbanti, L., Bonacina, C., Bonomi, G. and Ruggiu, D. 1972. Lago d'Orta: situazione attuale e previsioni sulla sua evoluzione in base ad alcune epotesi di intervento. ed. Istituto Italiano di Idrobiologia, Pallanza.

Bonacina, C. 1970. Il Lago d'Orta: ulteriore evoluzione della situazione chimica e della struttura della biocenosi planctonica. Mem. Ist. Ital. Idrobiol. 26:141-204.

Bonacina, C. and Bonomi, G. 1974. La conoscenza delle origini ed evoluzione dell'inquinamento del Lago d'Orta come base per la formulazione di strumenti previsionali indispensabili ad una politica di intervento. Atti 1° Congr. Internaz. Ambiente e Crisi Energia. Torino, 8-12 Maggio 1974: Vol. IV 25 pp. 13 figs.

Bonacina, C. and Bonomi, G. 1985. Advances in the recovery of lake Orta after three years of functioning of the new treatment plants. Proc. Internat. Congr. "Lakes Pollution and Recovery", Rome, 15-18 April 1985:95-100.

Bonacina, C., Bonomi, G. and Ruggiu, D. 1973. Reduction of the industrial pollution of Lake Orta (N. Italy): an attempt to evaluate its consequences. Mem. Ist. Ital. Idrobiol. 30:149-168.

Bonomi, G. and Adreani, L. 1978. Significato adattativo della struttura comunitaria e della dinamica di popolazione nel macrobenton profondo di un lago artifiale. In: Il Lago di Pietra del Pertusillo: definizione delle sue caratteristiche limno-ecologiche. ed. Istituto Italiano di Idrobiologia, Pallanza. 133-202.

Brewer, P.G. and Goldman, J.C. 1976. Alkalinity changes generated by phyto-plankton growth. Limnol. Oceanogr. 21:108-117.

Calamari, D. and Marchetti, R. 1975. Predicted and observed acute toxicity of copper and ammonia to rainbow trout (Salmo gairdneri Rich.). Prog. Wat. Techn.7:569-577.

Corbella, C. and Tonolli, V. 1958. I sedimenti del Lago d'Orta testimoni di una disastrosa polluzione cupro-ammoniacale. Mem. Ist. Ital. Idrobiol. 10:9-50.

De Agostini, G. 1927. Flora, pesca e fauna del Lago d'Orta. Cusiana 6(11):121-124.

Dillon, P.S., Yan, N.D., Scheider, W.A. and Conroy, N. 1979. Acid lakes in Ontario, Canada: characterization, extent and responses to base and nutrient additions. Arch. Hydrobiol. Beih. Ergebn. Limnol. 13: 317-336.

Gerletti, M. and Provini, A. 1978. Effect of nitrification in Orta Lake. Prog. Wat. Tech. 10:839-851.

Giaj-Levra, P. 1925. Diatomee del Lago d'Orta. Atti Soc. Ligustica Sci. Lett. 5.

Haux, C. and Larsson, A. 1984. Long-term sublethal physiological effects on rainbow trout, Salmo Gairdneri, during exposure to cadmium and after subsequent recovery. Aquatic Toxicology 5:129-142.

Ingersoll, C.G. and Winner, R.W. 1982. Effects on Daphnia pulex (De Geer) of daily pulse exposures to copper and cadmium. Environm. Toxicol. Chem. 1;321-327.

Monti, R. 1929. Limnologia comparata dei laghi insubrici. Verh. Internat. Verein. Limnol. 5:462-497.

Monti, R. 1930. La graduale estinzione della vita del limnobio del Lago d'Orta. Rend. Ist. Lomb. Sc. Lett. 63:3-22.

Mosello, R. Bonacina, C., Carollo, A., Libera, V. and Tartari, B. (in press). Acidification due to in-lake oxidation: an attempt to quantify the proton production in a highly polluted sub-alpine italian lake (Lake Orta). Mem. Ist. Ital. Idrobiol.

Muniz, I.P., Seip, H.M., and Sevaldrup, I.H. 1984. Relationship between fish populations and pH for lakes in southernmost Norway. Water, Air and Soil Pollut. 23:97-113.

Pavesi, P. 1879. Nuova serie di ricerche sulla fauna pelagica nei laghi italiani. Rend. Ist. Lomb. Sc. Lett. 63:3-22.

Provini, A and Gaggino, F. 1985. Metal contamination and historical record in Lake Orta sediments. Verh. Internat. Verein. Limnol., 22:2390-2393.

Schindler, D.W., Turner, M.A. and Hesslein, R.H. 1985. Acidification and alkalinization of lakes by experimental addition of nitrogen compounds. Biogeochemistry 1;117-133.

Stumm, W. and Morgan, J.J. 1981. Aquatic Chemistry. New York: John Wiley and Sons.

Vollenweider, R.A. 1963. Studi sulla situazione attuale del regime chimico e biological del Lago d'Orta. Mem. Ist. Ital. Idrobiol. 16:21-125.

Vollenweider, R.A. 1965. Materiali ed idee per una idrochimica delle acque insubriche. Mem. Ist. Ital. Idrobiol. 19:213-286.

Winner, R.W. 1984. The toxicity and bioaccumulation of cadmium and copper as affected by humic acid. Aquatic Toxicology 4:267-274.

Winner, R.W. 1976. Toxicity of copper to daphnids in reconstituted and natural waters. EPA-600/3-76-051.

Wright, R.F. 1977. Historical changes in the pH of 128 lakes in southern Norway and 130 lakes in southern Sweden over the period 1923-1976. SNSF project TN 34/77:71pp.

CHAPTER 7

RIVERINE AND OTHER TROPICAL LAKES AND THEIR FISHERIES

John E. Bardach

East-West Center and University of Hawaii at Manoa
1777 East-West Road, Honolulu, HI 96848

ABSTRACT

Discussion centers around large shallow lakes, mostly in the tropics and their flood plains. Such water bodies as the Great Lake of Cambodia, Lake Chad and others, as well as China's largest fresh water lakes have been and are being affected by changes in water regimes related to population increases and shifts. The resulting fairly rapid alterations in availability of water, spatially and temporally, as well as pollution have greatly reduced fish production in these waters. It has even led to the loss of certain species. In view of the importance of fresh water fisheries in the nutrition of many millions of riparian dwellers, development of measures to cope with these changes are of great urgency. Traditional and still ongoing fisheries in large shallow lakes were often based on extensive aquaculture techniques which now cannot be practiced any more. Intensification of aquaculture is being advocated to make up for shortfalls in annual fish harvests. Obstacles to the realization of such measures, here to be treated also, are technical as well as economic and social.

This presentation is dedicated to the memory of our colleague Karl E. Lagler.

INTRODUCTION

While the population of the North American Great Lakes region, like that of most others in industrialized nations, will barely grow between now and the mid-21st century (U.S. Dept. of Commerce 1979), the population of certain tropical nations in Asia and Africa will double in the same span of time (Pop. Ref. Bureau 1986). The significance of these projections for future staple and other food commodity requirements is that world grain, meat and fish demand is likely to increase by fifty percent in the next twenty years and more than double again in the next half of the next century. One should note in this regard that during the last decade, inland fisheries grew at a faster rate than those in the sea, with the catches from lakes, streams and ponds now making up nearly fifteen percent of aquatic harvests destined for direct human consumption (FAO 1984). Contrasting with this number, the ratio of fresh and salt water on the globe - the former is less than one percent by surface of the latter - underscores the importance of large and small lakes and their watersheds for the provision of animal protein. Implied here are both the greater average productivity of inland waters as opposed to oceans and the greater amount of control man could establish over the former as opposed to the latter. Lake and river productivities compared to those in the sea are more pronounced in the tropics - with which this paper mainly deals - than in the temperate zone. Relative productivity levels and importantly also, ease of control of access help to explain the striking difference in growth during the last decade between capture and culture fisheries. In the last decade aquaculture production grew at a yearly rate of seven to eight percent compared to a roughly two percent annual increase in tonnage of marine food fishes (Bardach 1985).

Naturally, questions arise about the quantity of land available in the world and about the productivity of the waters that flow over it to feed the population of the future. In inland fisheries, there is serious concern about soil erosion which connects the land on the watersheds with the rivers that flow through it and the lakes that lie on it. In India, for instance, 90 million hectares out of 328 million, or about twenty-six percent, are affected by water erosion while in the Argentine, the figure is thirteen percent of the total cultivated area (Hauck 1985). It is shallow rather than deeper lakes which are quickly and severely affected by increased run-off, usually also carrying high loads of fertile elements and not infrequently toxic wastes.

TABLE 1 NUTRITIONAL SIGNIFICANCE OF FISH AND SEAFOODS IN THE DIETS OF PEOPLE IN ASIAN COUNTRIES

Country	Per capita GNP US$ (1976)	Daily per capita availability (1972-74)				
		Calorie	Protein g	Protein of animal products g	Fish and seafoods g	In % of animal protein
East Asia						
China, People's Republic of	410	2,278	61.7	11.8	3.8	33.0
China, Taiwan	1,070	2,757	74.2	24.7	11.1	44.9
Korea, Republic of	670	2,749	72.0	13.1	9.0	68.7
Korea, DPR	470	2,636	76.4	12.1	8.0	66.1
Japan	4,910	2,832	85.5	40.1	22.1	55.1
Hong Kong	2,110	2,641	78.6	44.4	14.8	33.3
Mongolia	860	2,477	92.8	60.9	0.2	0.3
Southeast Asia						
Singapore	2,700	2,823	74.6	37.6	13.7	36.4
Malaysia, Peninsula	860	2,539	45.1	8.6	1.0	11.6
Malaysia, Sarawak	–	2,518	51.9	15.5	9.3	62.0
Malaysia, Sabah	–	2,776	59.7	22.1	11.3	51.1
Indonesia	240	2,031	42.0	5.3	3.6	67.9
Philippines	410	1,957	45.5	16.7	8.9	56.7
Thailand	380	2,302	49.5	13.2	6.8	51.5
Burma	120	2,125	55.2	8.8	4.0	45.5
Kampuchea	–	2,081	48.4	7.5	3.2	42.7
Laos	90	2,064	56.2	9.5	1.9	20.0
South Asia						
India	150	1,967	48.5	5.3	0.8	15.1
Bangladesh	110	1,949	43.2	6.7	3.5	52.2
Pakistan	170	2,128	57.2	12.7	0.3	2.3
Sri Lanka	200	2,075	40.9	6.6	2.6	39.4
Nepal	120	2,019	49.7	7.5	0.1	0.1
Afghanistan	160	2,001	61.5	6.9	–	–
Iran	1,930	2,319	60.8	11.8	0.1	0.1

Prepared by Y. H. Yang, RSI/EWC, from FAO (1977) and World Bank (n.d.).

The animal protein which inland waters can or should produce are of greater importance in the diet of a large portion of Asia than elsewhere (Table 1, Figure 1) but they also count in the nutrition of Africa. Therefore, it is pertinent to ask under what conditions the inland waters, especially lakes, of many rapidly growing developing nations can continue in their role as providers of this important human nutritional component? To treat this question, observations will be presented on changes in fish production of several large shallow lakes in developing nations.

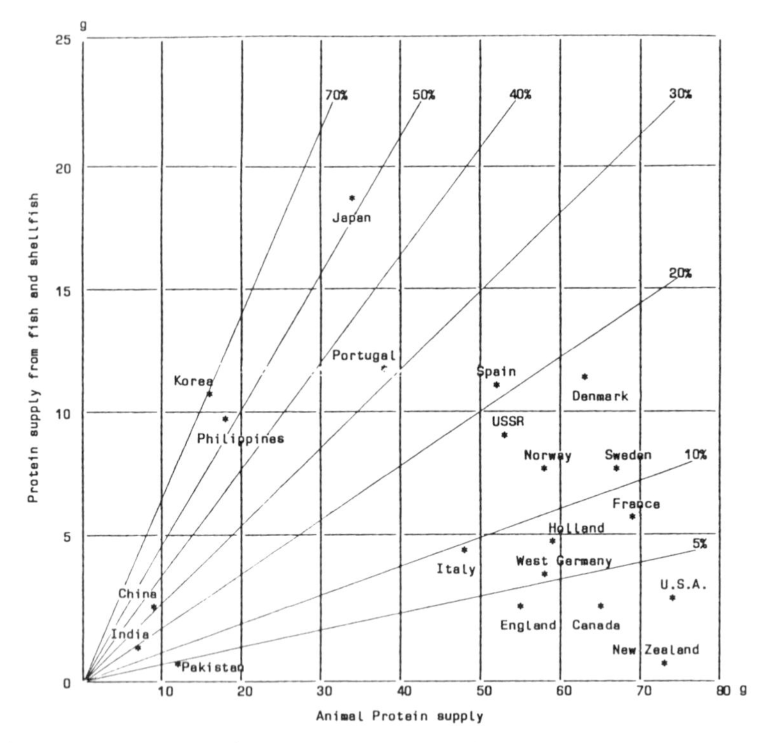

Figure 1. Protein supply from fish and shellfish in various countries (per person per day) (Japan Fisheries Assoc. 1980).

WATER REGIMES AND FISH PRODUCTION

In contrast to deep large trench or rift lakes such as Tanganyika, many larger tropical lakes are but a few meters deep. As they lie in climates with pronounced rainy seasons, their volumes and surfaces fluctuate greatly throughout the year. These are frequently in the courses of large river systems (Mekong, Nile, Hoang Ho) and four-fold or even greater differences between open water reaches at high and low levels have been recorded. The Great Lake of Cambodia, for instance, has a permanent water surface of 2,500 km^2 but floods to about 11,000 km^2 at the height of the rainy season in September and October (Welcomme 1979). Inundated regions were originally forested, slowing the flow of water and affording variegated feeding and breeding grounds and protection to the fishes. While that forest cover has mostly disappeared, or is in the process of doing so, around most tropical riverine lakes with large areas of inundation, the fishes still behave as if they had the scope to spread. After all, changes in land use and river modification by man proceed on an economic time scale while adjustments in animal behaviour, if this occurs, would take place in an evolutionary time frame.

The fishes of the inundation systems encompassing lakes and rivers behave remarkably similar throughout the tropics. They have been named either "black" or "white" fishes. (The terms, probably introduced by French biologists, refer in fact to the often darker or otherwise protective coloration in the first as against the lighter more silvery shades in the second of the two groups). Most of the black fishes move more or less laterally from lakes to inundated areas and return to the lake at low water, if they do not remain in residual pools and lagoons. A smaller number of such species can move moderate distances up or down river during high waters to favorite breeding places. Among the black fishes, there may be several "air-breathers" (clariid catfishes, lung fishes) and they are, in general, more tolerant of low oxygen and extreme temperature conditions than the so-called white fishes, which often engage in substantial spawning migrations or travels to low water refuges. These have been recorded for the Lake Chad-Yaeres, the Great Lake-Mekong and the Lake Victoria-Nzoia river systems, to name only a few (various authors in Welcomme 1979).

Needless to say, the modifications of rivers in which the lakes in question find themselves have reduced inundated areas, and therewith breeding and feeding grounds, if they have not actually eliminated entire lakes. When these changes are coupled with ongoing heavy increased fishing, the results can be devastating and it is usually impossible to clearly assign yield reduction or species disappearance to either one or the other factor.

The Mekong Great Lake Fishery declined from 70,000 metric tons (MT) in 1940 to 35,000 MT in 1967 (Committee for the Coordination of Investigations of the Lower Mekong Basin 1970), while _Probarbus jullieni_, a prized fish formerly found in this fishery, has now disappeared altogether (Sritingsook and Yoovetwatana 1977). Similarly, with changes in the vegetation cover in the basin of Lake Victoria, _Labeo victorianus_ greatly declined, even while the fishery for it and other species in the shore zone of the lake increased in intensity. (Cadwalladr 1965).

These examples are of pronounced deteriorations in Lake-river fisheries but reasonably large lakes may be also eliminated outright, as was the case in parts of China. With the irrational Maoist policy of "planting crops in the middle of lakes and on top of mountains", China's largest fresh water lake, Poyan Mu, shrank from 565,000 to 26,000 hectares, while the second largest lake, Dongting Hu, had its surface reduced by half through land reclamation. Hubei, the province of a thousand lakes, lost about 600 shallow riverine lakes of more than sixty hectares, while several larger lakes in the Hoang Ho system had their surfaces reduced by half or more between 1969 and 1978 (Smil 1984). The fisheries' effects of such drastic changes in land use may be gauged by noting that the natural inland catch in 1976 was estimated to have been only half of that in 1954, and that per capita fish consumption in formerly fish-rich municipalities went down in the aftermath of the cultural revolution. One should also note that much of the fish harvest so affected was actually from fish culture which was carried out in these lakes at varying levels of intensity, together with the capture fisheries. Since 1979, China's internal development policies have changed and special measures are being taken to reverse the deterioration of aquacultural and fishery sites.

FISH YIELDS, SEASONS AND TECHNIQUES

In 1981, the lakes and rivers of Asia and Africa furnished sixty (60) and seventeen (17) percent, respectively, of the total fresh water fish harvest of slightly over 8 million MT (FAO 1984). It is difficult to assign catches to lakes and river proportionately because of the already described contribution of inundated areas to fish production. Furthermore, the seasonal fluctuations in water levels lead to variations in fishing intensity by seasons (Figure 2) and expressions of yield over time may be difficult to compare. Even so, the relation of yield to inputs (labour and other energy and materials) over time appears highly rewarding in certain fishing installations used in many tropical lake-river complexes of the world (Marten and Polovina 1982). These so-called brush shelter or brush parks ("samras" in Cambodia, "acadjas" in West Africa, to give only two examples by local name) and well described by Welcomme and Kapetsky (1981), can be extensive and complex. They have evolved in artisanal fisheries over decades, if not centuries. They all have in common that they aggregate and shelter fish, thereby most likely increasing in them the already very high primary productivity of the shallow water. For harvesting, the installations are encircled by a net or fence, usually assisted by falling water levels; in some cases, they are also used as reservoirs for stocking other brush parks.

In contrast to yields from untended capture fisheries of a few to usually less than 100 kg/ha/yr, the yields from shelters are as high as those from more intensive aquaculture operations in ponds reaching in Africa 20 tons/ha/yr (Welcomme and Kapetsky 1981). The actual fishing season is far shorter though (Figure 2). Related to these shelters but representing higher fixed inputs, are net cages or pens, such as are found in Laguna de Bay, the largest lake of the Philippines, described by Pullin (1981).

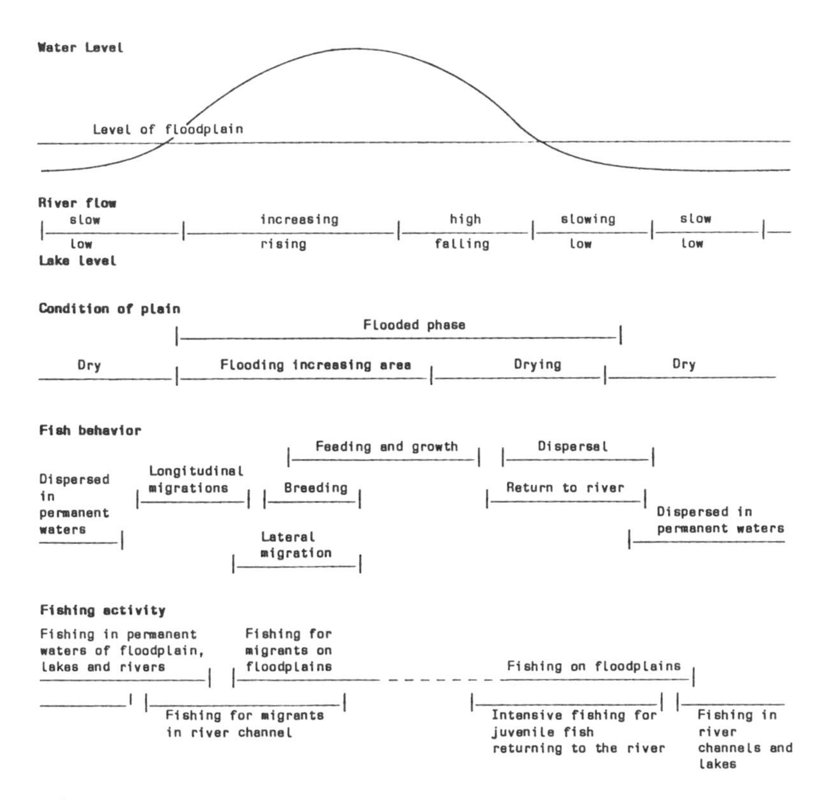

Figure 2. Summary of year-round conditions and activities on a floodplain and in shallow lakes (modified from Welcomme 1979).

The lake has a catchment of about 300,000 hectares and a surface of about 90,000 hectares. Its average depth is 3 meters. Its deepest spot is 10 meters at low water, and its waters are frequently stirred and therefore often turbid. In the winter, it tends to be slightly brackish. Since the mid-1960s, milkfish and tilapia (Seratherodon) pen fisheries have developed rapidly, the two kinds of fishes relying on filter feeding phytoplankton, although zooplankton, benthos, and filamenteous algae are also consumed. The dominant phytoplankton groups in the lake vary seasonally. During the summer, blue-green algae are usually dominant, but shifts between blue-green and green algae, diatoms and zooplankton abundance can occur throughout the year. This unpredictable and uncontrollable situation can pose problems. The lake can develop dense algae blooms, particularly of the blue-green alga Microcystis which may cause localized oxygen depletion with risk of fish kills. The other main feeding base in the lake is detrital, derived from dead plankton and other decomposing organic materials and having, in the past, produced large catches of snails, bottom-feeding fishes and other benthos. Tilapias filter plankton like milkfish, can also utilize bacterial protein and other nitrogen sources in detritus vary efficiently. The present large tilapia yields (below) claimed from pens probably derive from both plankton and detritus and may be be sustainable, especially if pens are crowded together. The use of fish pens in Laguna de Bay, especially the very large milkfish units, represent a form of ranching, in which fingerlings are released into a natural productive lake partitioned for easy harvesting.

With so many variables, it is all but impossible to formulate an optimum nutrient budget for the fish pen industry, which now seems to avail itself of some sixty (60) percent of the lake surface. This is deemed to overtax the natural productivity of the lake. In part, because of turbid conditions, dissolved nutrients in the water column are not being converted to biomass as rapidly as they should be in a tropical lakes (Nielssen 1983). It will be interesting to see whether pollution control (if implemented) and a proposed dam-gate complex have a smoothing or enhancing effect on plankton bloom and collapse cycles in the long-term. Another physical influence that threatens fish production is the proposed dredging, mainly to maintain lanes for small ship traffic. It will increase turbidity in places and most likely depress fish yields (Catalan 1983).

An interesting question is the future balance between tilapia and milkfish production. As things look, feeding versatility, resistance to low dissolved oxygen and increasing market acceptance favor the tilapia. The pens for tilapia are smaller than those for milkfish and give higher yields per unit surface. Four to seven tons of milkfish per hectare are reported while the tilapia harvests per hectare, albeit with supplemental feedings of rice bran, have been said to be as high as one (1) ton per L/Lo hectare in four to five months.

Brush park and net pan fisheries can have several problems. The lakes in which they are practiced are common domain and their productive capacities are easily overtaxed. Also brush uses for fish shelters and for firewood may compete with one another and the installations slow the flow of water and hasten siltation. Their management may pose questions of entitlement, as is the case in Laguna de Bay, where many tilapia growers are "squatters" who can be displaced by the original landowners. In view of the rapid changes in land use and intensification of riparian agriculture, artisanal practitioners may fare badly compared to larger entrepreneurs, and where the lake site is close to a city, like Metro-Manila in the case of Laguna de Bay, or an industry, serious pollution problems can arise.

CHALLENGES TO MANAGEMENT

Certain fisheries of very large tropical lakes can stand intensification but these are mostly in the open reaches of a few deep lakes of Africa, with their partly unexploited stock of small schooling planktivores. In large reservoirs, where stocks of selected species have been introduced, increases in fishing pressure may also be possible (Petr and Kapetsky 1983). By and large, though, human population increases, coupled with the scaling up of inputs into agriculture, especially through irrigation, lead to modifications of flood plans, their rivers, and shallow lakes. In the process of aiming for various benefits, such as flood control, hydropower and irrigation, multiple, often conflicting uses of water are intensified. The shore lands, once access paths to the fisheries or sites of extensive culture operations, cannot fulfill these functions any more.

With ongoing changes in water distribution of tropical lake-river complexes, there is hardly an author concerned about national or regional fish supplies who does not advocate intensification of aquaculture

(Welcomme 1979, Micha 1981, as examples). When one examines the prerequisites for greatly enhancing the development of aquaculture under the conditions in question, one must note that there are physical, economic, bio-technical and socio-cultural factors to contend with in the execution of such schemes.

Among the physical factors, chemical pollution tends to assume a special place and centers around unwise use of chemical pest control in intensified agriculture. Food chain effects of organic pesticides are now well established and some of their mechanism, such as hepatoxicity and reproductive failures, are elucidated. Mirex and PCBs (Aroclor), for instance, are implicated in suppression of vitellogenin production induced by estradiol upon low-level, long-term exposure (Chen et al. 1986).

Even though the pesticide problem is highly complex, regulations, while not wholly satisfactory and prone to breaches, are at least reasonably well in place in developed nations. Developing countries could, in theory, build on similar patterns of pesticide use that would not expose their environments and their public health to serious harm. But apparently they did not take the necessary measures because policy and decision makers in low income countries are skeptical about the costs involved in implementing environmental protection laws and regulations which had been found effective elsewhere. They feared that environmental impact assessments as practiced, say under the U.S. National Environmental Policy Act, were impractical for local conditions with few technically trained people and would take overly long to execute and would slow, if not delay, needed, especially agro-technical, changes (Carpenter and Dixon 1985). Their fears are probably justified and there are no easy answers to this general dilemma.

To make matters worse, their world governments unfortunately also subsidize pesticide application. Production and sales are favored through access to foreign exchange on favorable terms, tax exemptions or reduced rates, easy credit and sales below cost by government-controlled distributors. Studies into the matter in countries in Africa, Asia and Latin America showed farmers will tend to adopt more chemical intensive strategies than are economically justified when there are pesticide subsidies. These, to quote from Repetto (1985):

> ... "encourage farmers to use chemicals with serious environmental side effects, and undermine efforts to promote integrated pest

> management techniques. Also they slow down research to develop biological controls and reduce changes to develop past monitoring networks and to better enforce existing pesticide regulations."...

In most countries where irrigation agriculture is on the rise, and with it pesticide use, the just-mentioned programs rarely exist or are weak and inadequate (Carpenter and Dixon 1985). Sublethal action of pesticides will continue to affect fish, wildlife, domestic stock and possibly human beings. Not all locations are equally affected though. Thus it is of primary importance to assess the pesticide levels in the waters destined for aquaculture and to establish hatcheries with uncontaminated water supplies.

Other physical concerns are space and water. The patterns of their availability differ in different river systems and lower courses of tropical rivers, even without permanent lacustrine formations, and become extensive seasonal shallow lakes of high waters. The Lower Gangae River in Bangladesh, with one of the largest flood plains in the world, is a case in point (Figures 3 and 4) (Aquatic Farms Ltd. 1986):

> At the height of the rainy season, as much as 52,000 km^2 or over one-third of the country is under water. After the monsoon, these flood waters recede into rivers, beels, baors and haors (oxbow lakes and low lying depressions) covering a (dry season) area of well over 1,000,000 hectares.
>
> The floodplain along with the rivers, beels, baors, haors and lakes have traditionally been the major source of fisheries production. In 1983/84, these waters yielded total fisheries production of 465,000 MT, sixty-one percent of the total national output from all sources.
>
> Average daily animal protein supply per capita in Bangladesh is about 5 grams, only about 2/3 of the level of protein supply in other south Asian countries and far below that in Southeast Asia. Apparent fish consumption averages 7.5 kg annually per capita, supplying over fifty percent of animal protein (Table 1).
>
> Although the floodplain fishery is an important source of animal protein,

Figure 3. Satellite photo of northcentral Bangladesh on November 10, 1984 showing the floodplain management areas I & II (darkest areas indicate water). MA-I (from Sunamganj to Brahmanbaria) still has about 450,000 ha inundated on this date. MA-II can be seen as the two dark prongs extending north of Dhaka.

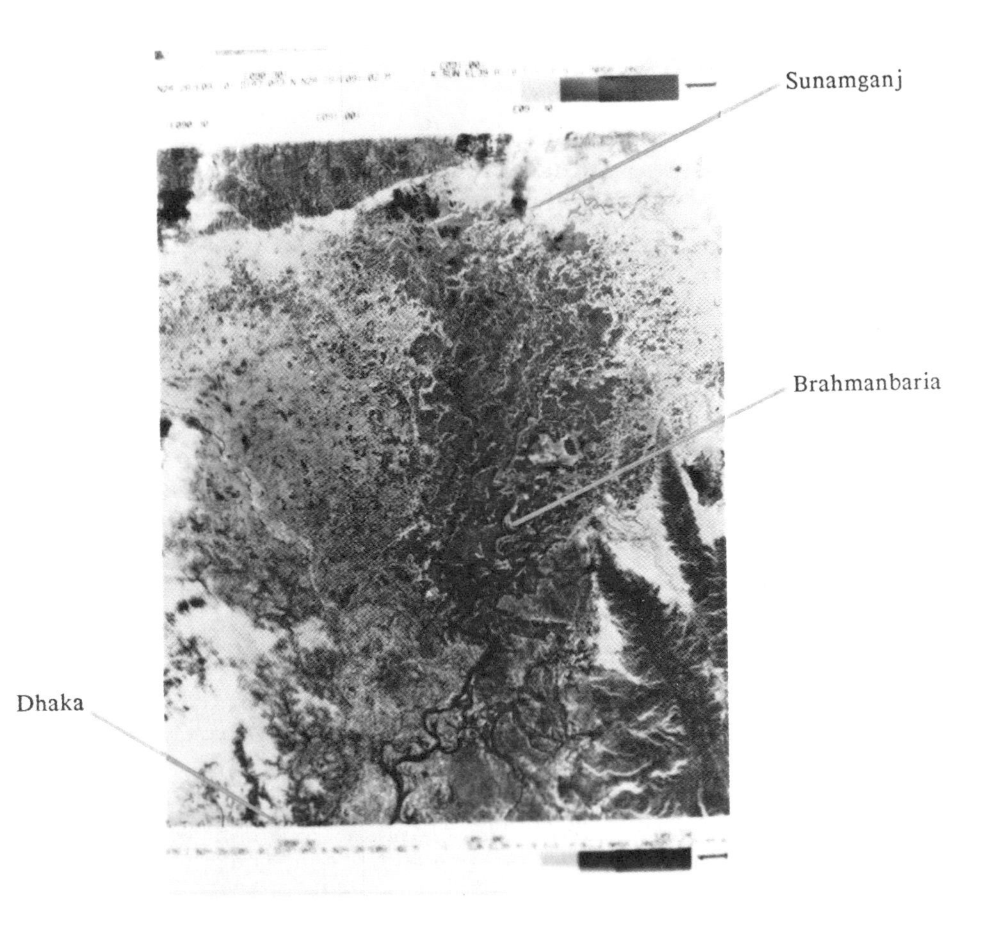

Figure 4. March 1985 satellite photo showing the same area of northcentral Bangladesh as pictured in Figure 3. Note that the water (darkest area) has receded into rivers and beels.

> substantial efforts have been made in recent years to control floods because of the damaging effect on agricultural crops, property and human life. Over 150 flood control projects have been initiated and many more are planned. It is anticipated that by the year 2000, the net negative impact of these projects, on the natural fishery, largely to benefit agriculture, will be a decline in fish production of 150,000-250,000 MT annually.
>
> The rich agricultural lands that are flooded each year provide excellent conditions for fish. Although flood control projects will continue to reduce the size of the floodplain, there still will be an enormous productive area within the foreseeable future. By the year 2000, the floodplain should be well in excess of 2 million hectares during the monsoon. Its waters will continue to be rich in nutrients, with the chief limiting factor in their productivity being the lack of juvenile fish to utilize these nutrients. If it is possible to manage the fishery in such a way as to expand the fish population to fill the habitat to its full natural carrying capacity within the four to six month flood period, the annual fresh water capture harvest could be increased by a factor of three times or more. While no attempt has previously been made to artificially stock the floodplain, this strategy appears to offer the best single hope of producing the large quantities of fish required to feed the country's growing population.
>
> In 1985, the country's hatchery production of hatchlings of major and Chinese carp was just over 1 billion. The present total hatchery capacity in the country is approximately 8-9 billion when all developing hatcheries are in full production. To fully exploit the floodplain, tens of billions of fry could be required.

If more intensive pond development were to be made of making good fish shortfalls due to river regulation, finding pond sites may become important. Take the case of the Great Lake of Cambodia-Mekong complex for instance, where about 35,000 MT of natural fish harvest were lost until fifteen years ago, when statistics were still available (Committee for the Coordination of

Investigations of the Lower Mekong Basin 1970). If one were to make good this shortfall by semi-intensive pond culture, albeit with low energy and material inputs, and based on recycling of nutrients, such as is achieved by combining fish with poultry or pig rearing, annual harvests of three metric tons per hectare may well be attained (Bardach 1981, 1985). The number of hectares needed would then be 12,000 or 600 farms of twenty hectares of water surface each. Even if the farms were smaller, their yield a bit lower, and their numbers were closer to three thousand, the physical characteristics of the Cambodian plain would permit such a development and more. Inasmuch as lake-river complexes discussed here generally occur in lower river courses with much flat land, purely physical obstacles to a partial transition from fisheries to aquaculture would not be paramount. The recovery of the lake and river fisheries of China, largely through intensification of aquaculture, is a case in point (Smil 1984). If pen culture in stabilized lakes is increased, selection of such sites must be made with attention to ecological and limnologic variations in the lake so as to avoid both overstocking and excessive euthrophication (Nielssen 1983). Questions of entitlement to the installations also become important.

Nor would the obstacles lie in selection of suitable species. Polyculture complexes of Chinese carp, Indian carp, or their combinations, and also tilapia as well as _Clarias_, are capable of yielding high tonnages per hectare of pond or reservoir surface with known and proven low to medium intensity management techniques (Micha 1981) and application of genetics promise to lead to even higher yields (Maclean 1984). Adaptation to culture of several suitable species of _Colossoma_ is under way in Latin America (Saint-Paul 1985). These techniques would have to include the presence of hatcheries in many cases and vigorous pursuit of location-adapted extension efforts. The sensitivity of hatcheries to pollution has already been mentioned but it should be added that they are also sensitive to economies of scale, the latter varying within the region which they are meant to support. Hatchery development and extension schemes should go hand in hand.

Aquatic farms also have economies of scale, varying in intensity with species location and socio-economic conditions. In certain locations in the Congo, for instance, it was demonstrated that a small fish farm of 1,500 to 2,000 m^2 (Tilapia) would bring sustaining revenues to a manually laboring farmer. If better income is desired and a truck were, for instance, to be added to the farm machinery, a fish farm of four hectares is still in deficit whatever the management regime. In order to attain an interesting profit with management that includes more mechanization, the water surface should measure at least 10 hectares (Micha 1981) and in the more "developed" economic climate of Hawaii, an area of about 20 hectares is the bare minimum to make a Malasian prawn farm succeed (Shang 1981). As already intimated, these differences can be accentuated by the comparisons of only roughly comparable situations. In the case at hand, Malasian Prawn (Macrobrachium rosenbergii), though bringing a high return per unit weight, is a bottom dweller and a somewhat territorial one at that, in contrast to the fish which utilized the entire water column, permitting one to attain higher yields in weight per unit surface.

Commitment by governments to the development or upgrading of aquaculture requires that the activity be included in the range of land uses for the support of which regulations are revised or established. First and foremost among them is to ensure to the aquafarmer security of land tenure and entitlement to water use. Aquatic animal husbandry cannot be effectively pursued when the entrepreneur - be he owning, leasing or even squatting - cannot look forward to the amortization of his inputs.

That these require capital is self-evident. The pay-out of funds for aquaculture development, be it for government hatcheries or for the installations of private entrepreneurs, is the end of a long chain of events. Like in other areas of agricultural development, the chain starts in national plans with the seeking of low interest loans from international financing institutions. After a series of fact-finding missions to assess technical and financial feasibilities, which may or may not include socio-cultural considerations, the funds find their ways through national into agricultural banks or their equivalents. Funds then move to those who would tend and rear aquatic organisms, where they previously only gathered them. There is as yet limited experience on the social forces at work in rural transformations, such as the change-over from fisheries to aquaculture, from slash and burn farming to agro-forestry and from growing floating rice to irrigation rice culture. They are all of a kind and those involved in them, prominently including government personnel who supposedly are to facilitate the changes, must re-think some very basic concepts and re-fashion their attitudes, not only towards the meaning of work and prevailing interpersonal relations, but also towards land and water ownership and long-term responsibilities of land and water conservation, as examples (Roy 1984).

While technical instruments for changes are usually known in broad outlines and often in detail, their successful implementation at the grassroots level is more difficult to achieve because social and cultural factors are difficult to take into account and because of the very speed with which population increase, urbanization, technical and economic development act on traditional societies.

CONCLUSIONS

Population increases in tropical countries, leading to more intensive land use and urbanization, bring in their wake the modification of large river systems and of the lakes that lie in them. These have supplied the dwellers on and near their floodplains with an important fish component in their diet. The altered water regimes of modified rivers tend to reduce natural fish production even while satisfying new and more varied water uses than those which existed before modification. Pollution and soil erosion as well as over-fishing add their deleterious effects. While there are no satisfactory data to estimate the amount of these reductions, one can venture the guess that several hundred thousand metric tons could be involved world-wide.

Annually available quantities of water in the total system are usually not diminished by river modification but the pattern of its distribution is altered. This pattern would, however, permit increased water use for aquaculture in newly constructed ponds and/or enclosures in stabilized lakes and in such reservoirs as may have become part of the regulated river system. Techniques developed largely in Asia permit attainment of high fish production per unit water surface, even with relatively moderate inputs in energy and materials, especially when sound nutrient recycling systems are employed.

Main obstacles to, or forces working against, successful establishment of aquaculture are the overtaxing of natural carrying capacities in lakes, pollution and competition for uses of the land, especially in the vicinity of cities. Serious also, in the medium long run, appears to be the competition for water. Coastal aquaculture is already growing faster in several countries than that in fresh water. The latter, however, retains an edge through the relative ease of control one can establish over the infrastructural system of the culture installations. Mollusc and macro-algae culture, of course, are exceptions to this statement.

Aquaculture of various intensity levels in stabilized lakes and on the flood plains of modified tropical rivers has the technical potential of more than making up for the losses that occurred in the fisheries of the lake-river systems in questions, and of increasingly satisfying regional demands for fish. Realizing that potential will, however, depend on changes in various conditions that are extraneous to aquaculture biotechnology proper. Of great importance among them are the institution, in many places, of vigorous pollution control, of equitable land and water uses and the planning for infrastructures that can help bring about changes in primary occupations, help people learn the new skills, adjust to altered modes of income generation, etc., etc. These lie in the socio-economic realm of the transformations that accompany rapid development, and without attending to them, the technical ones rarely come to full and proper fruition.

REFERENCES

Aquatic Farms Ltd. 2/1986. Bangladesh Second Aquaculture Development Project Preparation Report. Final Report. Prepared for the Government of the People's Republic of Bangladesh and the Asian Development Bank. Honolulu, Hawaii. 71 pp. 17 appendages.

Bardach, J.E. 1985. The role of aquaculture in human nutrition. Geo. Journal 10(3):221-232.

Bay of Bengal program. In: Development of Small Scale Fisheries, Madras, India. FAO.

Cadwalladr, D.A. 1965. The decline in the Labeo victorianus Blgr. (Pisces: Cyprinidae) fishery of Lake Victoria and an associated deterioration in some indigenous fishing methods in the Nzoia River, Kenya. East Africa Agric. & For. J. (3)249-256.

Carpenter, R.A. and Dixon, J.A. 1985. Ecology meets economics: a guide to sustainable development. Environment 27(5):6-11, 27-35.

Catalan, Z.B. 1983. Dredging the Laguna de Bay. PESAM Bulletin, Sept.-Oct., pp.3, 9, Los Banos, Laguna, Philippines.

Chan, T.T., Reid, P.C., Sonstegard, R.A. and van Beneden, R. 1986. Effect of aroclor 1254 and mirex on estradiol-induced vitellogenin production in juvenile rainbow trout (Salmo gairdneri). Can. J. Fish & Agric. Sci. 43:196-173.

Committee for the Coordination of Investigations of the Lower Mekong Basin 1970. Report on indicative basin planning: Proposed framework for the development of water and related resources Lower Mekong Basin. UN/ECAFE, Bangkok.

FAO (Food and Agriculture of the United Nations) 1984. FAO Yearbook of Fisheries Statistics. 56. Rome.

Hauck, F.W. 1985. Soil erosion and its control in developing countries. In: Soil Erosion and Conservation. eds. S.A. El-Swaify, W.C. Moldenhauer, and A. Lo, pp. 718-728. Ankeny Iowa: Soil Conservation Society of America.

MacLean, J. 1984. Tilapia - the aquatic chicken. ICLARM Newsletter 7(1):3-17, Int'l Ctr. for Living Aquaculture Res. Man., Manila, Phillippines.

Marten, G.G. and Polovina, J.J. 1982. A comparative study of fish yields from various tropical ecosystems. In: Theory and Management of Tropical Fisheries. eds. Pauly, D. and Murphy, G.I. pp. 255-285, ICLARM/CSIRO, Manila.

Micha, J.C. 1981. Aquaculture. Potentialities actualles et futures en eaux douces. Bull. Francais de Pisciculture 284:178-188.

Nielssen, B.H. 1983. Limnological studies in Laguna de Bay. Terminal Report to SEAFDEC Aquaculture Department, Manila. Philippines. (Mimeogr. 58 pp.)

Petr, T. and Kapetsky, J.M. 1983. Pelagic fish and fisheries of tropical and subtropical natural lakes and reservoirs. ICLARM Newsletter. 6(3):9-11.

Population Reference Bureau, Inc. 1986 World Population Data Sheet. Washington, D.C.

Pullin, R.S.V. 1981. Fish pens of Laguna de Bay, Philippines. ICLARM Newsletter. 4(4):11-13.

Repetto, R. 1985. Pesticide subsidies in developing countries. World Resource Institute, Research Report Number 2, pp. 1-27, Washington, D.C.

Roy, R.N., ed. 1984. Consultation on Social Feasibility of Coastal Aquaculture. 125 pp.

Saint-Paul, U. University of Hamburg 1985. Potential for aquaculture of South America freshwater fishes. Manuscript submitted to Aquaculture.

Shang, Y.C. 1981. Comparison of rearing costs and returns of selected herbivorous, omnivorous and carnivorous aquatic species. Marine Fisheries Rev. pp. 43-49.

Smil, V. 1984. The Bad Earth Environmental Degradation in China. 245 pp. M.E. Sharpe, Inc. (Armonk, New York).

Sritingsook, C. and Yoovetwatana, T. 1977. Induced spawning the Pla Yee-Sok (Probarbus jullieni, Sauvage). In: Proceedings of the Indo-Pacific Fisheries Council. Sec. I and II, 17th Session. SYM 37. Bangkok: FAO Reg. Off. Asia and Far East, 1977.

U.S. Department of Commerce, Bureau of the Census 1979. Illustrative Projections of State Populations by Age Race and Sex 1975 - 2000. Washington D.C.

Welcomme, R.L. 1979. Fisheries Ecology of Floodplain Rivers. Longman, London and New York. 317 pp.

Welcomme, R.L. and Kapetsky, J.R. 1981. Acadjas: The Brush Park fisheries of Benin, West Africa. ICLARM Newsletter 4(4):3-4.

CHAPTER 8

LAKE BIWA, ITS VALUE AND RELATION TO TRIBUTARY STREAMS

Syuiti Mori

Head Office, Shiga University, Hikone, Japan

ABSTRACT

Lake Biwa is located at the central part of Japan and although the surface area is small when compared to the Great Lakes it is the largest lake in Japan. Lake Biwa has diverse value for human life: the water maintains the lives of 13 million people and sustains the industrial works of the second largest industrial area in Japan. Large amounts of fishes, molluscs and shrimp are produced, much of which are endemic to this lake. The scenery around this lake is one of the most aesthetically pleasing in Japan. Lake Biwa has an extraordinarily long history, between 2 and 3 million years, which indicates that this lake is the second or third oldest lake in the world and that it has been formed by repetition of land depressions and deposition of gravels or soil carried from surroundings through tributary streams. Owing to this long history some organisms have developed characteristic habits, such as the fish Ayu or crucian carps. More than 460 streams are flowing into this lake and the main effects of these streams are:

1. carrying gravels or soil, and

2. carrying various contaminants, including nutrients, herbicides and heavy metals, to the lake.

As for the nutrients, about 8,320 kg of total N and 878 kg of total P are conveyed daily to the lake, principally by streams. According to the prohibiting rules harmful influences from insecticides have remarkably decreased, but there is no prohibitive rule for herbicides as their effects should be noticeable. Heavy metals are uniformly deposited in the bottom sediments but no real injury is observed.

INTRODUCTION

Lake Biwa is the most famous, important and heavily investigated lake in Japan. The real features seem not to be well known to the people of the world even when they are scientists (Figure 1). First, the author wishes to explain briefly the value of this lake from various viewpoints, and secondly, touch on some aspects of the characteristic features of organisms living there which have developed with the remarkably long history of this lake. Thirdly, the relation of tributary streams to this lake will be discussed from the point of carrying:

1. gravels and soil, and

2. contaminants such as nutrients and toxic substances - herbicides and heavy metals.

VALUE OF LAKE BIWA

Some important statistics concerning Lake Biwa are:

Catchment Area		3,848 km^2
Lake Surface Area		675 km^2
Depth		deepest 104 m; average 40 m
Lake Volume		27.5 x 10^9 m^3
Coast Line		240 km
Inflowing Rivers and Streams		more than 460
Outflowing Rivers and Streams		4
		(River Seta is most important with 100 m^3/s)
Average discharge		about170 m^3/s
Transp.(1984)[1]:	N. Lake	5.2-7.4 m (ave. 6 m)
	S. Lake	1.5-2.2 m (ave. 1.8 m)
Nutrients(1984)[1] (mg/L) Tot.N:	N. Lake	0.16-0.47 (ave. 0.27)
	S. Lake	0.17-0.78 (ave. 0.39)
Tot.P:	N. Lake	0.004-0.027(ave. 0.007)
	S. Lake	0.006-0.085(ave. 0.023)

[1] Environment White Paper of Shiga Prefecture, 1985

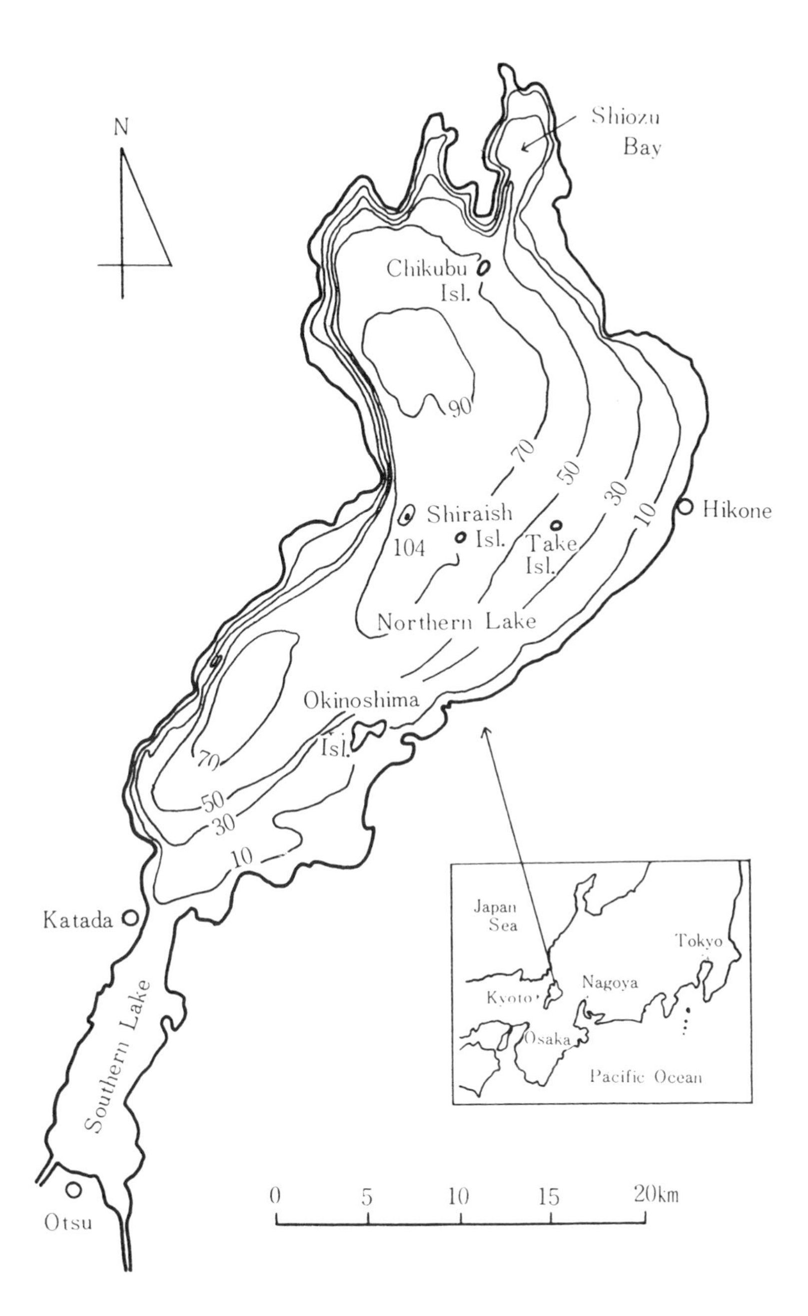

Figure 1. Lake Biwa with depth contours (Mori 1980).

VALUE OF LAKE BIWA FROM VARIOUS STANDPOINTS

Lake Biwa has various values not only for human life but also for its environment

1.The lake serves as the water supply for the 13 million people living in the Kinki District (about 11% of the total population of Japan).

2.The Kinki District is the second largest industrial area in Japan. The water from Lake Biwa sustains the manufacturing works distributed in this area.

3.Lake Biwa produces a large quantity of fisheries products, about 4.2 millions kg (7.4 billion yen) in 1984. This production is ranked as the fourth in quantity and first in yield among all Japanese lakes. Main products are fishes, molluscs and shrimps. One-third of the yield is made up by the fish Ayu, one-third is attributable to the production of freshwater pearl, and the remainder due to the production of all other fishes, molluscs and shrimps.

4.The present state of Lake Biwa seems to have been formed at least 2 million years ago. This fact was recently ascertained by Horie et al. (unpublished) by using air guns and drilling, as shown in Figure 2. Base rock was clearly found at depths greater than 900 m where the geological age was estimated to be about 3 million years. This history shows Lake Biwa is the second or third oldest lake in the world (the oldest lake is Lake Baikal, whose geological history is estimated at 15 million years). Thus the existence of Lake Biwa undoubtedly furnishes a very valuable store of scientific knowledge.

5.Many kinds of endemic species inhabit this lake. For example, 40 species of molluscs are living in this lake, among which 18 species are endemic. Thinking that about 60 species of freshwater molluscs are found in Japan, about one-third of the total species are found only in this lake (Mori and Miura 1980). Similar phenomena are found for other species such as fishes.

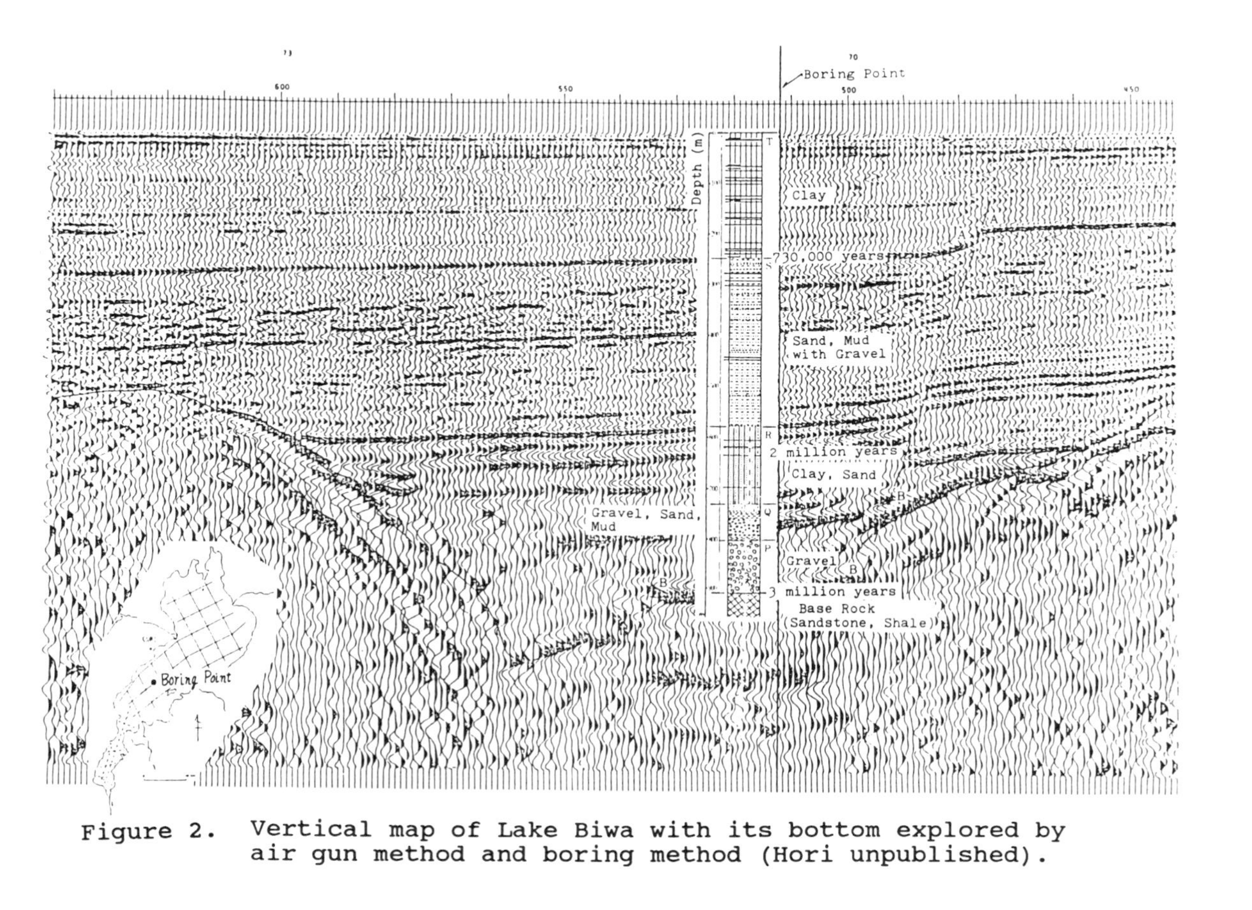

Figure 2. Vertical map of Lake Biwa with its bottom explored by air gun method and boring method (Hori unpublished).

These phenomena seem to be caused be a remarkably old geological history of this lake together with a good supply of various kinds of habitats for organisms living in this lake.

In some cases we can retrace evolutionary routes, along which endemic species developed from original common species, throughout a long geological history. This also demonstrates the incomparable value of this lake. For example, three species of crucian carp live in this lake, i.e. Carassius auratus langsdorfi (Cuvier et Valenciennes), Car. auratus grandoculis (Tem. et Schl.) and Car. auratus cuvieri (Tem. et Schl.). Among these, the first species is a common species found all over Japan and the other two are endemic to this lake. The evolutionary route is considered to have progressed from the first species to the last, from living benthic life and eating benthos to living pelagic life and eating phytoplankton. (Tomoda 1962, 1965, Suzuki 1974). A similar phenomenon is also found in the catfish group. Parasilurus asotus Lin. is a common species distributed widely in Japan. From this species two endemic species are considered to have been developed in this lake, that is, Par. lithophilus Tomoda and Par. biwaensis Tomoda. The former is specially fond of living in stony places. The latter, growing to as large as 1 m., lives in pelagic areas similar to a shark in the sea. (Tomoda 1962b).

6.The scenery around Lake Biwa is one of the most famous attractive in Japan. Its value is aesthetic. This has always been true throughout the long historical era of Japanese civilization. Many resort places for swimming, sailing, fishing, etc. can be found there.

VARIOUS ROLES OF TRIBUTARY RIVERS AND STREAMS

More than 460 rivers and streams flow into the lake. They have played various roles throughout geological time. The author wishes here to examine these roles from two sides:

1. the development of characteristic fish, and

2. the transportation of various substances, such as gravels, sand or soil, and contaminants such as nutrients, herbicides and heavy metals.

DEVELOPMENT OF CHARACTERISTIC FISH - AYU

As mentioned earlier, the fish Ayu, <u>Plecoglossus altivelis</u> (Tem. et Schl.), is the most important commercial fish produced in Lake Biwa. It represents one-third of the total yield from this lake. While this fish is widely distributed all over Japan, in Lake Biwa it is developing a few unique characteristics.

Normally when rivers flow into the sea, the Ayu eggs are spawned near the river mouth in the autumn. The hatched fry swim to the sea where they spend winter near the coast area. When spring comes they approach the river mouth and swim up the river until they reach the middle of the river where they rapidly grow to adulthood during summer. When autumn comes they swim down to the river mouth to spawn and then die.

However, on the other hand, the Ayu in Lake Biwa behaves a little different. The Ayu spawns eggs as usual near the mouth of the river which flows into the lake and the hatched fry swim into the lake. They then winter in the lake as if they were living in the sea and in the next spring they approach the river mouth to swim upstream. Here they divide into two groups.

One group, smaller in number, swims upstream to the middle of the river and the other, larger group does not enter the river but remains in the lake throughout the Ayu's whole life. The Ayu of the former group eat benthic algae and grow up to adult size as large as 20 cm in body length. In autumn they swim down to the river mouth for spawning. These are known as catadromous fish.

The latter group eat mainly zooplankton throughout their life and only attain a body length of about 8-12 cm even at the adult stage. When autumn comes they come to the river mouth, swim up only a short distances, spawn and die. In this case they are anadromous fish.

This difference in body length between two life modes seems to be caused by the lack of food of the lake and alternately, the plentiful food in the rivers. When small individual Ayu are caught in the lake and stocked in rivers they grow rapidly to the normal adult size.

Recently a problem related to the lowering of the Lake Biwa water level has emerged. A large scale plan to exploit water from this lake is in progress. When this plan is executed, it is estimated that the water level will fall by as much as 1.5 m. Under these circumstances, the river mouths will dry up and Ayu that remain in the lake will not be able to reach spawning sites in autumn. This will exert a remarkably harmful influence on the life of the Ayu and at the same time on the fisheries in this lake.

In order to prevent this harmful influence, the construction of an artificial stream for spawning was attempted. Figure 3 shows one of these streams. The water is supplied to this stream by pumping up lake water. Thus the flowing water is always secured at the spawning site in the mouth of the stream. It is expected that 7 billion fry will be produced in these artificial streams and will be deposited in the lake in autumn. It is also expected that 70 million Ayu (300,000 kg) will be caught when they grow up to commercial size. This amount is nearly equal to the current natural yield.

TRANSPORTATION OF MATTER BY TRIBUTARY RIVERS AND STREAMS

Various substances are transported to the lake by tributary rivers and streams. Among these, the author wishes to mention:

1. gravels, sand and soil, and

2. contaminants such as nutrients, herbicides and heavy metals.

Transport of Gravels, Sand and Soil

The long history of Lake Biwa is indicated in the repetition of depressions of land and depositions of gravels, sand and soil (Figure 2). As a result, tremendous amounts of gravels, sand and soil have accumulated on the bottom of Lake Biwa. These materials must have been transported through millions

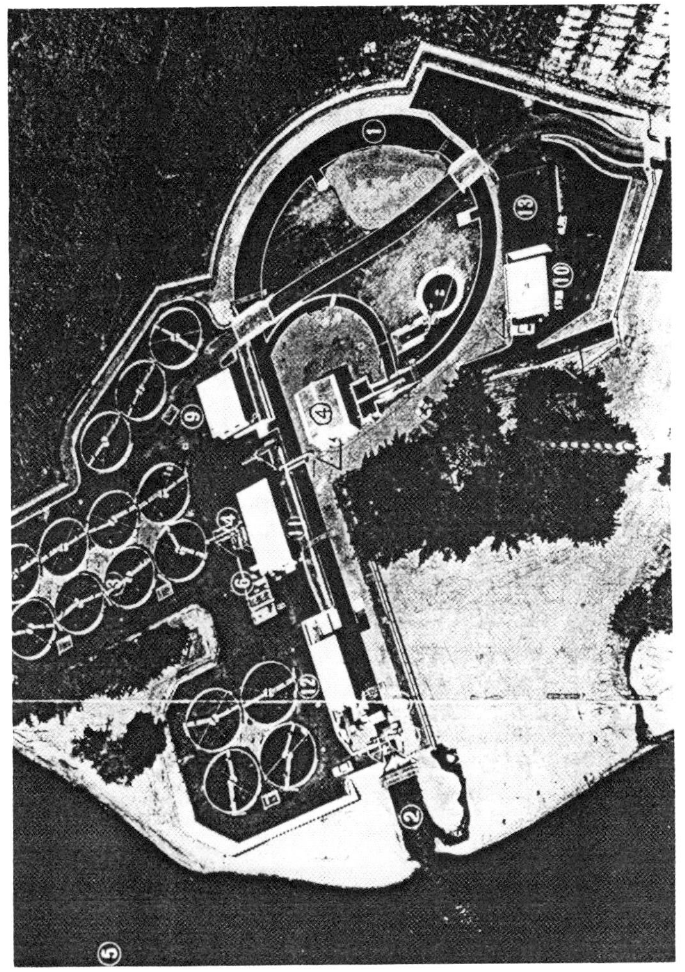

Figure 3. Artificial stream to provide spawning site for the Ayu. 1. spawning site; 2. mouth of the stream to the lake; 4. facilities for pumping up groundwater; 5. Lake Biwa. Small circular ponds seen above are rearing ponds for the Ayu (from Fisheries Experiment Station of Shiga Prefecture).

of years by tributary rivers and streams. Unfortunately we have no accurate data at present on how much of these materials are transported to the lake yearly. However, we can make a trial calculation if we assume that the particulates deposited yearly on the bottom of the lake are those transported by tributary rivers and streams. We can, then, calculate the amount of transported particulates by measuring the depth of yearly deposition on the lake bottom.

It is widely recognized that the depth of yearly deposits is 1-2 mm at the central part of the lake. This means that 700,000 - 1,400,000 m^3 of particulates are deposited yearly on the bottom of the whole lake. The majority of which must have been transported by tributary rivers and streams.

As a case history, the changes of the river mouth of the River Kusatsu over a 6 year period are introduced in Figure 4. From March of 1979 to May of 1985 the coast line near the river mouth advanced into the lake about 62.5 m by depositing sand carried by the River Kusatsu. This river is famous for carrying a great quantity of sand. Due to deposition the river bed has risen. Today, the main road and even railroad (Tokaido Line) are lower than the river bed.

It should be noted that irrespective of the considerable deposition of gravels, sand and soil, the depth of the lake as a whole seems to have remained nearly constant over a fairly long time. This seems to indicate that a depression of land is compensating for the effect of deposition.

Transport of Contaminants

Nutrients

Sometimes nutrients are not contaminants for lakes but in the case of Lake Biwa, which is mesotrophic or even oligotrophic in some parts, they are usually considered as pollutants.

Table 1 shows the gross nutrient (total N and total P) budgets of Lake Biwa. Data of "inflow" and "outflow"

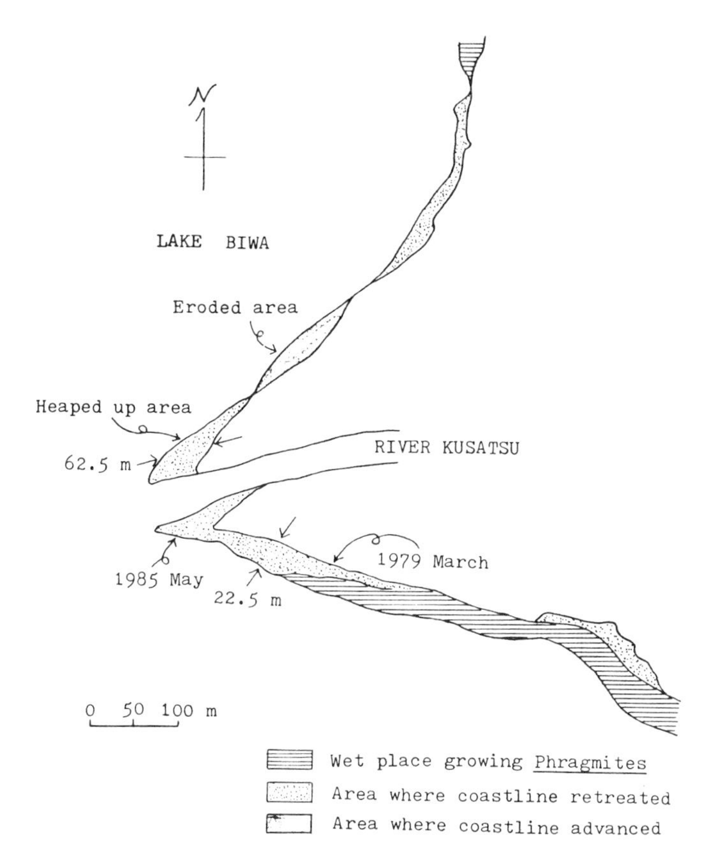

Figure 4. Advancement of land into Lake Biwa at the Kusatsu River mouth over a 6 years period (by air photo) (Kumagai et al. 1985).

TABLE 1 GROSS NUTRIENT BUDGETS IN LAKE BIWA

	INFLOW (kg/day)		OUTFLOW (kg/day)	
	Total N	Total P	Total N	Total P
Rivers[a]	8,320	878	6,360	497
Precipitation and Falling Dusts	2,900	96		
TOTAL	11,220	974	6,360	497

[a] Inflow - Data from 133 rivers and streams
Outflow- Data from 4 outlets
(Calculated from the data of Kunimatsu 1981)

were calculated by using the survey data obtained by Kunimatsu (1981) on 133 rivers and streams and 4 outlets, respectively. About 8,320 kg of total N and 878 kg of total P are carried daily into the lake. In addition about 2,900 kg of total N and 96 kg of total P are brought daily into the lake through precipitation and dust fall. In total 11,220 kg of total N and 974 kg of total P are added daily to the lake. On the other hand, 6,360 kg of total N and 497 kg of total P are flowing daily from the lake through 4 outlets. A remarkably large amount of nutrients remain in Lake Biwa, 4,860 kg total nitrogen and 477 kg total phosphorus. This represents about 43% and 49% of the added nutrients, respectively. A fraction of these nutrients are circulated in the water column while the remainder is deposited on the lake bottom. This means that the trophic condition is more or less continuously enriched, a common fate of freshwater bodies.

Based on various data, it must be concluded that a drastic change in the lake organism world had occurred in the early 1960s. The main cause should be attributed to the progress of eutrophication. The phenomenon was first observed on the benthos. The oligochaetes biomass at the central part of the northern lake increased rapidly after 1962, from 5 g/m^2 in 1962 to 27 g/m^2 in 1972. This was maintained until 1976 whereupon it decreased to 8 g/m^2 in 1980. On the other hand, the number of cells counted in 1969, 400/m^2, increased continuously to 1700 in 1980. The discrepancy in figures between weight and number of cells is due to the alteration of living species, i.e. during this period Branchyura Soverbyi Beddard, which is heavy, decreased in number and Tubifex hattai Nomura and Limnodrilus sp., which are light, increased in numbers enormously (Mori 1980, 1984, Narita unpublished).

A similar observation can be made of a kind of mussel, the Corbicula group. Corbicula sandai Reinhardt (one of the endemic species) had thrived in the southern lake up to early 1960s with a density of about 90 g/m^2, but it then decreased and almost disappeared by 1974. Instead, Corbicula leana Prime, which is a common dweller in the freshwater of Japan and is more resistant to pollution, appeared in early 1970s and took hold. It attained a density of 64 g/m^2 in 1974. However, even this species nearly disappeared in 1977 (Mori 1984). Of course over-fishing is somewhat responsible for this phenomenon, but it depends mainly on the excess nutrients in water - eutrophication.

A little later than the appearance of drastic changes in the world of benthic organisms, there appeared also a conspicuous change in the world of planktonic organisms. In 1977 the red tide caused by Uroglena americana (Calkins) Lemmermann was first observed in the northern lake. Later, though, showing a yearly fluctuation, the appearance of the phenomenon of red tide occurs regularly in early summer, even today (Kadota 1984).

Herbicides

Serious harm to fishes and other organisms by some pesticides such as PCB, BHC, DDT and others have been experienced. However, at present, their uses are prohibited by law. Therefore, although some amounts of DDT or BHC remain in the bodies of fishes or other organisms, they are decreasing year by year (Watanabe and Ishida 1983).

On the other hand, some kinds of herbicides are still used in the cultivated fields and they are flowing down through tributary rivers and streams to the lake. Among which CNP (4-nitrophenyl 2,4,6-trichlorophenyl ether) has been studied in detail by Nakaminami et al. (1985). Figure 5 shows the contents of CNP in rivers and the lake: a clear decrease of the concentration of CNP in water from rivers to lakes can be observed. The change of CNP content in organisms was studied by Ishida (1982) with the fish Isaza, Chaenogobius isaza Tanaka. The result is shown in Figure 6. CNP has appeared in the body of the Isaza since 1970, peaked in 1976 and then decreased somewhat. The real toxicity of CNP is not accurately ascertained but as it is reported that a strong toxic substance, dioxin (polychlorinated dibenzo-p-dioxin, PCDD) though it is a very small amount, is contained in CNP. Attention must be paid to this herbicide, which should be studied more precisely hereafter.

Heavy metals

Heavy metals found in the surface sediments of Lake Biwa have been brought into the lake from two sources. One is from a mine and the other is from disposable wastes originated in human and industrial activities. The heavy metals, of course, have been brought into the lake by tributary rivers and streams. The old copper mine was located at the north-eastern watershed of Lake Biwa. It had operated for 65 years since 1901 and heavy metals such as Cr, Cu and Cd are densely distributed in the surface sediment of the north-

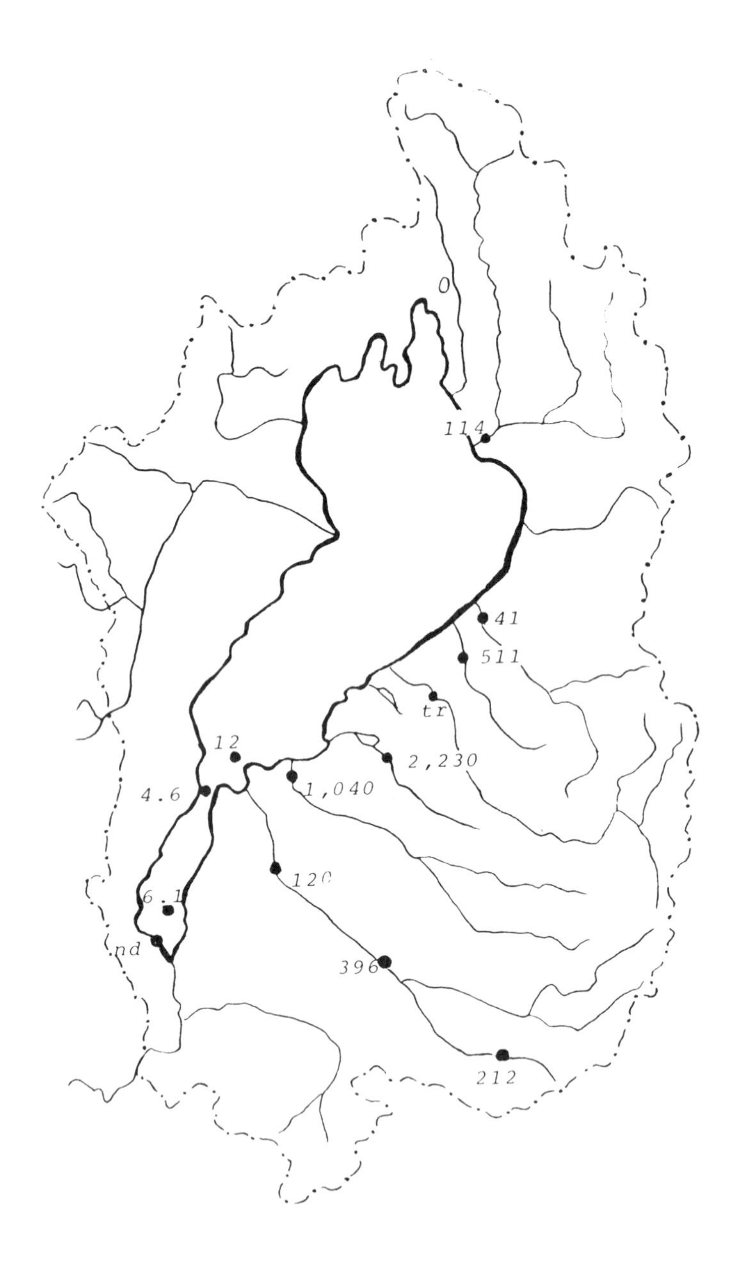

Figure 5. Contamination of Lake Biwa and some of the tributary streams by CNP in May 1984 (Nakaminami et al. 1985).

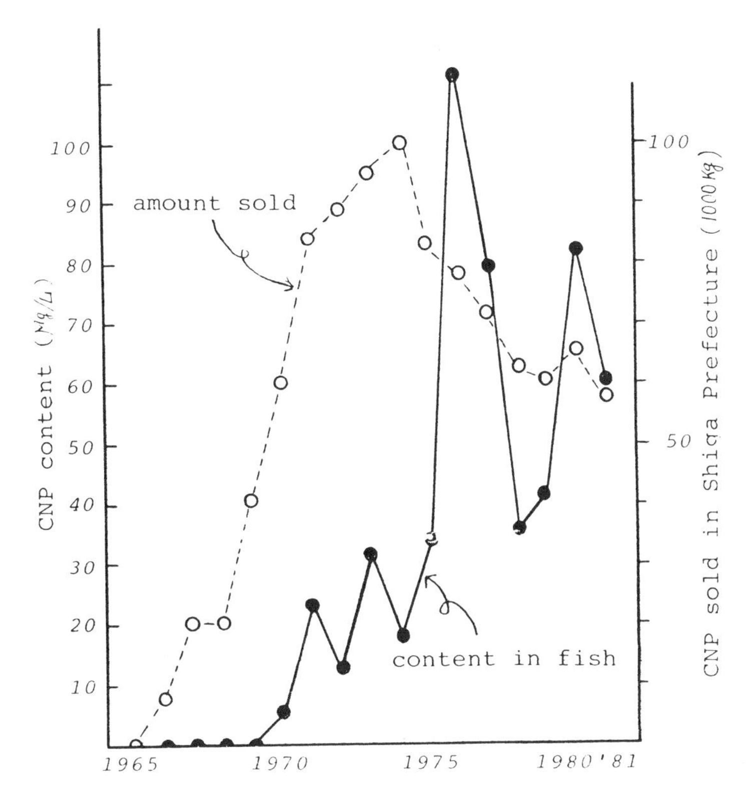

Figure 6. CNP content in Isaza (<u>Chaenogobius</u> <u>isaza</u>) and the amount of CNP sold in the Shiga Prefecture (Ishida 1982).

eastern part of Lake Biwa. On the other hand, such metals as Co or Pb are largely found in the surface sediment of the southern part of the lake. These metals seem to have entered the lake from the densely populated and industrialized areas around the southern part of the lake (Figure 7) (Tatekawa 1980). Although the heavy metals are distributed in the surface sediments of the lake as mentioned above, no serious injury has ever been reported.

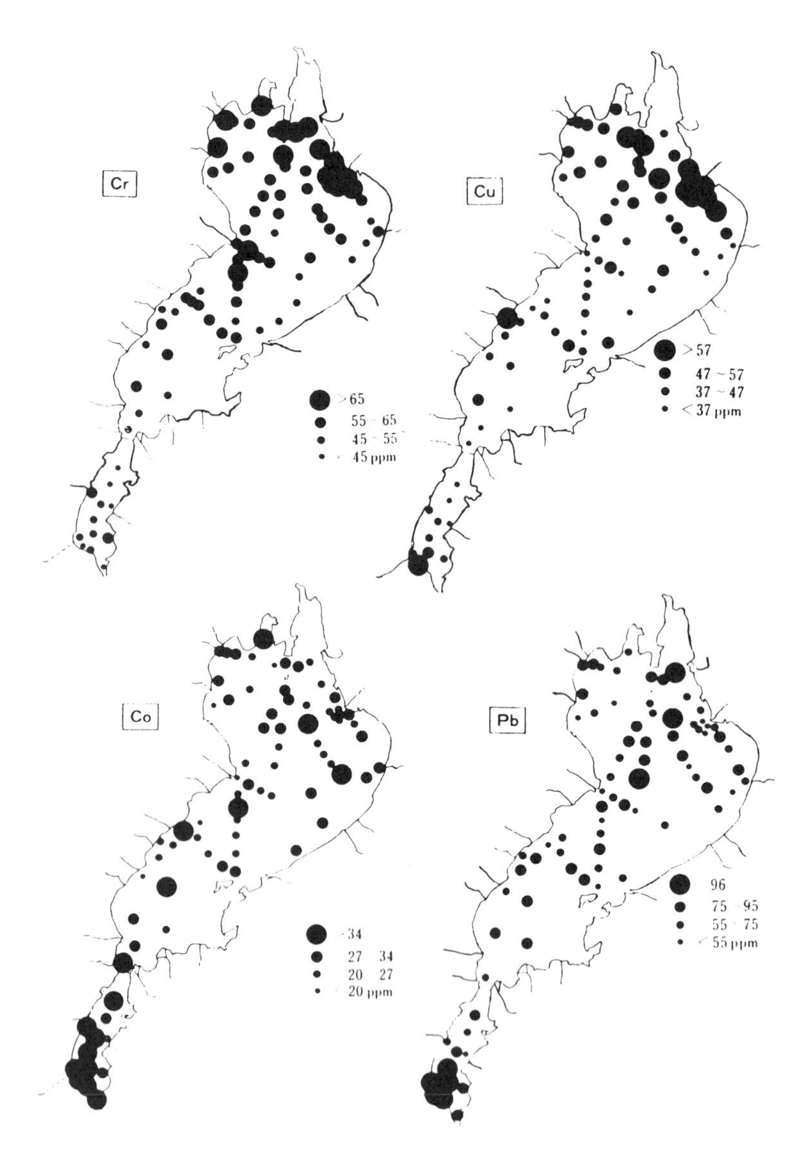

Figure 7. Some heavy metals in the surface sediments of Lake Biwa. Cu and Cr are mainly found in the sediments in the north-east and Co and Pb are principally distributed in the south (Tatekawa 1980).

REFERENCES

Horie, S. ed. 1984. Lake Biwa. Dordrecht, Dr. W. Junk Publishers.

Ishida, N. 1982. Environmental pollution by herbicide CNP. News from the Consumer Movement. No. 159:5-7 (in Japanese)

Kadota, H. 1984. The outbreak and its background of so-called freshwater red tide in Lake Biwa. Interdis. Res. Inst. Envir. Sci. 3:161-168 (in Japanese)

Kumagai, M., Maeda, H. and Onishi, Y. 1985. Estimation of influence due to lowering of the water level by using the remote sensing method. In: The report of the survey on The influence due to extraordinary water shortage of Lake Biwa in 1984. Lake Biwa Research Institute, pp. 147-174 (in Japanese)

Kunimatsu, T. 1981. Water qualities of river water, underground water and precipitation in the watershed of Lake Biwa. In: Environmental dynamics of Lake Biwa and its watershed. Report of 1980. ed. Special Research Group of Environmental Science, Ministry of Education, Culture and Science. pp. 82-104 (in Japanese).

Mori, S. ed. 1980. An introduction to limnology of Lake Biwa. Kyoto.

Mori, S. 1984. Ecosystems of Lake Biwa. Interdis. Res. Inst. Envir. Sci. 3:151-160 (in Japanese).

Mori, S. and Miura, T. 1980. List of plant and animal species living in Lake Biwa. Mem. Fac. Sci., Kyoto Univ. Ser. Biol. 8:1-33.

Nakaminami, G., Ishida, N., Kunimatsu, T. and Tsuruyama, S. 1985. The pollution of Lake Biwa, Yodo River, Chikugo River, Ariake Sea and others by diphenyl ether herbicide. Ecol. Chem. 8:3-10 (in Japanese).

Shiga Prefecture. 1985. White Paper on the Environment (in Japanese).

Suzuki, N. 1974. Respiratory responses to hypoxic conditions in crucian carp living in different habitats. Bull. Jap. Soc. Scientific Fisher. 40:57-62 (in Japanese).

Tatekawa, M. 1980. Heavy metals in the surface sediments. In: An introduction to limnology of Lake Biwa. ed. Mori, S. pp. 35-41.

Tomoda, T. 1962a. On some fishes of Lake Biwa. Biwa-ko Research. 3:38-45 (in Japanese).

Tomoda, T. 1962b. Studies of the fishes of Lake Biwa-ko 1. Morphological study of the three species of catfish in the genus Parasilurus from Lake Biwa-ko, with reference to their life cycle. Jap. J. Ichthyol. 8:126-146 (in Japanese).

Tomoda, T. 1965. Morphological studies on the development of the crucian carp in Lake Biwa. 1. Ontogenetic development of Nigoro-buna and some remarks on the differentiation of Gengoro-buna and Nigoro-buna. Bull Osaka Museum Nat. Hist. 18:3-29 (in Japanese).

Watanabe, N. and Ishida, N. 1983. Pursuit of residues of organochlorine substances in Lake Biwa. Ecol. Chem. 6:28-36 (in Japanese).

CHAPTER 9

EFFECTS OF CONTAMINANTS LOADINGS ON FISHERIES YIELDS FROM LARGE LAKES

R.A. Ryder

Fisheries Branch, Ontario Ministry of Natural Resources, Thunder Bay, Ontario, P7B 5E7

The intent of this short introduction is to provide an entry into the subject of contaminants effects on fish yields, a topic that is somewhat enigmatic in terms of cause-effect relationships. Most available evidence demonstrating the effects of contaminants on community fish yields is circumstantial in nature and rarely takes into account the multiple interactions set in motion once the most sensitive species are affected. Neither do single-species bioassays provide more than a slight indication of what to expect in a multi-species assemblage subjected to contaminants.

In this discussion of contaminants, the author has narrowed the scope to include only those toxic or potentially toxic substances either augmented or introduced into the aquatic environment of large lakes that may have direct or indirect toxic effects on fishes and thereby influence fish levels. Accordingly, putrescible materials will not be considered herein.

Toxic contaminants may be of two fundamental types (Ryder and Edwards 1985). The first type is "natural" in the sense that the substance occurs commonly in nature, although normally at relatively low concentrations. These natural materials, though toxic at high concentrations, may be a metabolic necessity for fishes at trace levels. Many heavy metals fall into this category. Because fishes have evolved over eons in the presence of heavy metals, they have developed certain inherent adaptations to them at normal, ambient concentrations. The toxic effects of heavy metals from anthropogenic sources, therefore, are primarily those of increasing quantities, as concentrations augment beyond the genetic capability of the fish to adapt metabolically. The second category of contaminants consists of "aberrant" substances created by man. Fishes have no evolutionary adaptation to these recently synthesized, xenobiotic materials. Toxic effects of aberrant materials may be disproportionately great even at low concentration levels. The ultimate effect on fishes is more of a qualitative nature, therefore, than those emanating from natural substances. The subsequent effect on fish yields, however, may be indistinguishable from that of high levels of "natural" contaminants.

Generally, toxic effects of both natural and aberrant contaminants range from non-significant at extremely low concentrations to marked toxicity at high concentrations. In the case of high levels of toxic concentrations, most fishes would succumb, result into close to one hundred percent mortality, as would most other gill-breathing organisms, including both vertebrates and invertebrates. Only certain plants or bacteria would likely survive. In the latter instance, those bacterial forms dependent on the contaminant as an important energy resource may even thrive at levels toxic to most other biota (e.g. Warren 1971).

The sub-lethal effects of toxic contaminants on a fish community, and therefore, on fishery yields, are subtle, and insidious and virtually intractable from a management point of view. At very low levels, contaminants may even be seen to enhance fishery yields (Table 1), albeit only on a temporary basis if levels continue to increase. This seemingly "positive" response to low levels of toxicity results from increased mortality of eggs and fry in nursery areas exposed to contaminants. Survivors moving out of these vulnerable area find a superabundance of food due to reduced competition within the cohort, resulting in their increased growth and perhaps greater ultimate size (e.g. Ryan and Harvey 1977). Nonetheless, this fleeting effect may be interpreted by the judiciary as "no effect", rather than as a grave portend of events to come, a more likely interpretation of aquatic ecologists.

Rehabilitation measures may be most easily and economically put into place when the ecosystem is subjected to only low levels of contamination. However, if the sublethal concentrations are not perceived as a future threat to the desired levels of

TABLE 1 A GENERALIZED SCENARIO OF BIOLOGICAL AND SOCIO-ECONOMIC EFFECTS ON AN AQUATIC COMMUNITY AND A FISHERY, FOLLOWING A LINEAR INCREASE (1-5) IN CONTAMINANTS MIXTURES.

	Biological Effects	Socio-Economic Effects
1.	First detection in water column at trace levels. No biological effect evident.	No effect.
2.	Reduces recruitment at egg or larval stages of sensitive species or through reduced fecundity. Remainder of recruited cohort may grow faster and to larger ultimate size through compensatory, density-dependent mechanisms. Subtle genetic effects may be in place.	Fishery yields may be temporarily enhanced by provision of larger mean size of commercially desirable species.
3.	Recruitment of sensitive species may fail completely. Community compensatory mechanisms swing dominance to less sensitive species.	Fishery yields may be enhanced or degraded depending on the economic value of the persistent species. Need for rehabilitation not yet generally perceived.
	Species comprising the food web may be affected directly or indirectly causing debilitation or loss of critical components. These effects, in turn, may be transported up the trophic ladder, affecting fishes through their nutritional requirements or metabolically, through bioaccumulation. Genetic effects become apparent. First evidence of increased incidence of deformities or disease.	Higher levels of contaminants may be sufficient to stimulate regulatory agencies to issue advisaries against selling or eating affected fish. Moratorium on commercial and/or sport fisheries may be imposed resulting in economic and recreational losses. Rehabilitation prospects tractable. Bioassay studies may be initiated.
4.	Situation is exacerbated by synergistic effects, heavy metal mobilization and bioconcentration of organic contaminants. Biotic community disaggregates and species integration is lost. High mortality levels become the norm. Survivors are usually small pelagic species.	Complete loss of a valuable resource to human use. Small pelagic species of no commercial value persist. Rehabilitation becomes prohibitively expensive.
5.	No reproduction of any fish species possible due to high contaminants loadings. Remaining standing stocks lost by attrition over time. Incidence of deformities and/or disease high among remaining stocks.	Rehabilitation not feasible as long as contaminants loadings remain at high levels.

Table 1 A generalized scenario of biological and socio-economic effects on an aquatic community and a fishery, following a linear increase (1-5) in contaminants mixtures.

fishery yields, it is unlikely that appropriate mitigation or rehabilitation measures would be taken, or even considered!

At higher levels of toxic waste inputs, rehabilitative measures would likely be considered to be a suitable management option, particularly at the stage where fishes become unfit for human consumption. At these levels (Table 1), rehabilitation measures if at all feasible, will be more difficult and costly to implement. Benefit-cost ratios between potential fisheries yields and rehabilitative measures may not be easily justified by management agencies, particularly if a clear-cut cause and effect relationship is requisite in advance of legislative approbation.

The current state of our knowledge of cause-effect relationships between contaminants loadings and multi-species fish yields is abysmally deficient in most areas. Particularly distressing is our lack of understanding of how contaminants mixtures affect interrelationships within aquatic communities comprised of fishes and other aquatic organisms. Contaminant effects on food web linkages are still poorly understood. Future research efforts should be directed towards monitoring a variety of contaminant mixtures at the "interface zones": substrate-water, atmosphere-water, nearshore littoral, river deltas and other ecotones where contaminants are likely to have their greatest effects on aquatic communities. Concomitant research should be directed towards in-situ studies of contaminants mixtures on aquatic communities and their sublethal effects over time, including the food web effects and other interactions. Greater use of indicator organisms as surrogates of ecosystems (e.g. Ryder and Edwards 1985) in studies of contaminants effects would accelerate our accumulation of knowledge and provide more timely and appropriate management.

ACKNOWLEDGEMENTS

This is contribution No. 86-07 of the Ontario Ministry of Natural Resources, Research Section, Fisheries Branch, Box 50, Maple, Ontario, Canada.

REFERENCES

Ryan, P.M. and H.H. Harvey 1977. Growth of rock bass, *Ambloplites rupestris*, in relation to the morphoedaphic index as an indicator of an environmental stress. J. Fish. Res. Bd. Can. 34:2079-2088.

Ryder, R.A. and C.J. Edwards (eds.) 1985. A conceptual approach for the application of biological indicators of ecosystem quality in the Great Lakes Basin. Int. Joint comm., Windsor, Ontario. 169 p.

Warren, C.E. 1971. *Biology and water pollution control*. W.B. Saunders Co., Toronto, Ontario.

CHAPTER 10

INDEXES FOR ASSESSING FISH BIOMASS AND YIELD IN RESERVOIRS

Robert M. Jenkins

Aquatic Ecosystems Analysts, P.O. Box 4188, Fayetteville, Arkansas USA 72702

ABSTRACT

Studies of long-term trends in lake fish populations aimed at identifying the superimposed effects of contaminants are very difficult and expensive. Analyses are required for all processes controlling species population size, including pollutants that may be stressing the population. Improved quantitative monitoring methods are needed to establish cause and effect relationships. In the interim, simpler methods of estimating fish biomass and yield (Ryder et al. 1974) are available to provide comparative indexes to "normal" lake conditions and to aid in identifying perturbed ecosystems. Empirical multiple regression equations derived from analyses of data assembled from reservoirs >200 ha in the United States are now available for use as assessment indexes (Ploskey et al. in press). Variables include 23 physiochemical attributes of the environment and angler use and sport fish yield statistics from 380 reservoirs in 46 states. Biomass data by species, based on cove sampling, were assembled from 393 reservoirs in 26 states. Subsamples of these data sets, sorted by similar water chemistry, operational use, and surface area characteristics, were also analyzed. The regressions derived are not guaranteed to express cause-effect relations but do afford standards for comparison and quantitative predictive values. Until more precise natural lake ecosystem models are developed, it is recommended that analogous data sets, assembled on a drainage basin, regional, national or global basis, be analyzed in a similar manner to provide standards for sorting unstressed, naturally-stressed and contaminant-stressed communities. Such endeavors could help in the development of legally-adequate ecosystem epidemiology.

INTRODUCTION

The overall objective stated in most U.S. water pollution control laws and regulations has been to restore and maintain the physical, chemical and biological integrity of the Nation's waters. "Biological integrity" has been described as the basic properties of living aquatic communities:

1. the standing crop or abundance of organisms;
2. kinds and relative abundance of organisms, and
3. community metabolism (physiological) and condition (disease, histopathology and parasitism).

Ergo, these are the primary properties to measure to determine if pollution control objectives are met. In this paper an attempt will be made to address economical approaches to measurement of the first two properties in lakes.

Although the relative toxicity of contaminants can be measured accurately in the laboratory, the data are of limited value in predicting effluent effects on the biota because existing environmental conditions are usually poorly known, and aquatic ecosystems are complex and dynamic. Environments also undergo large cyclic or unexpected seasonal changes which alter the susceptibility of aquatic organisms to toxins.

An increasingly expressed view holds that approaching aquatic toxicology by lab experimentation alone is inadequate to define chronic low level exposure and we must start from observations of anomalous biological phenomena in the environment. In pleading for the development of epidemiology for chemically induced diseases in fish in Canada, Gilbertson (1984) concluded:

> ..."It is no longer enough to document exposure and experimental effects; we now have been forced to formally demonstrate the effects in the environment. Only after we have recognized a biological anomaly and formed an hypothesis can we return to the laboratory to undertake toxicological research and investigate its etiology, and to bring about controls that are based on rigorous science rather than polemics."...

Data are being laboriously acquired from field studies to permit better understanding of aquatic ecosystems, and efforts to model perturbations of fish populations have increased. However, broad extrapolation of field and laboratory measurements, combined with educated guesses, are required to construct detailed, compartment-type models. In addition to toxicants, stresses that may perturb large lake and reservoir fish populations include variations in water exchange rate, fluctuation in surface levels, excessive nutrient loading, over exploitation, invasion of new fish species and aquatic plants, shoreline modifications, water withdrawal for irrigation or industrial purposes, and natural eutrophication. The effects of these natural and man-induced stresses must be sorted out before the chronic, sublethal effects of toxicants can be specifically assessed.

Until more precise models are developed, it is suggested that simpler methods of estimating fish biomass and yield be used to provide comparative indexes to desirable lake conditions and help identify contaminant-perturbed environments. Empirical models derived by regressing fish yield and standing crop data on important environmental factors in United States reservoirs are available as examples of this general concept (Ploskey et al. in press).

The potential for using both abiotic and biotic variables as predictors of fish yield or biomass has been recognized for over 50 years (Welch 1935). Research in this vein has accelerated since the appearance of the morphoedaphic index (total dissolved solids/mean depth; Ryder 1965), and great progress has been made in the precision of predictive techniques and in understanding their limitations (Ryder 1982). Multiple regression analyses have provided formulae which enable biologists to easily predict fish production and harvest potential. The goal has been one of scientific parsimony (to explain as much as possible with as little as possible). If natural variations in fish standing crop and yield can be explained by differences in a few key environmental factors, then regional "standards" could be established to aid in identifying contaminant-stressed populations.

The approach recommended here is not new. For example, the morphoedaphic index was used, in conjunction with growth rate of fishes, as an indicator of environmental stress by Ryan and Harvey (1977). They studied the effects of acid rain on fishes in 24 La Cloche Mountain lakes, Ontario, and concluded that the index and growth rates appear to have considerable diagnostic value for lakes undergoing stress.

In using the morphoedaphic index or other predictive regressions at the global levels, climate and lake size have overriding effects. Indexes have their greatest utility at the regional level where growing season may be considered as a constant and area reduced to constant units. Here, variations in nutrient loading assume greater importance, together with lake morphometry which governs nutrient movement and available caloric heat (Ryder 1982).

U.S. RESERVOIR DATA BASES

This conference is addressing problems of large lakes on a global scale. How large is a "large" lake? Fels and Kellar (1973) proposed that a world register of man-made lakes should include those >10,000 ha or a mean depth >10 m . There are 78 U.S. reservoirs in their "large" area category, covering 1.9 million ha at mean annual pool levels. In comparison, excluding the Great Lakes and Alaska, there are 43 natural lakes >10,000 ha totaling 1.8 million ha. Bue (1963) defined the "principal" natural lakes of the United States as those >2,471 ha in area. The World Register of Dams (Brown et al. 1964, with supplements through 1968) lists, 12,000 dams >15 meters in height. Martin and Hanson (1966) included 1,562 reservoirs with a storage capacity of >56.6 hm^3 in a United States Geological Survey (USGS) nationwide inventory. This minimum capacity criterion corresponds to a surface area of about 202 ha the definition used here in acquiring information on U.S. man-made lakes.

Environmental data bases have been assembled by Aquatic Ecosystem Analysts which contain 23 variables representing descriptive, physical, and chemical characteristics of U.S. reservoirs with surface areas of $\geq$202 hectares. The variables included are measured routinely and related to the design, operation,

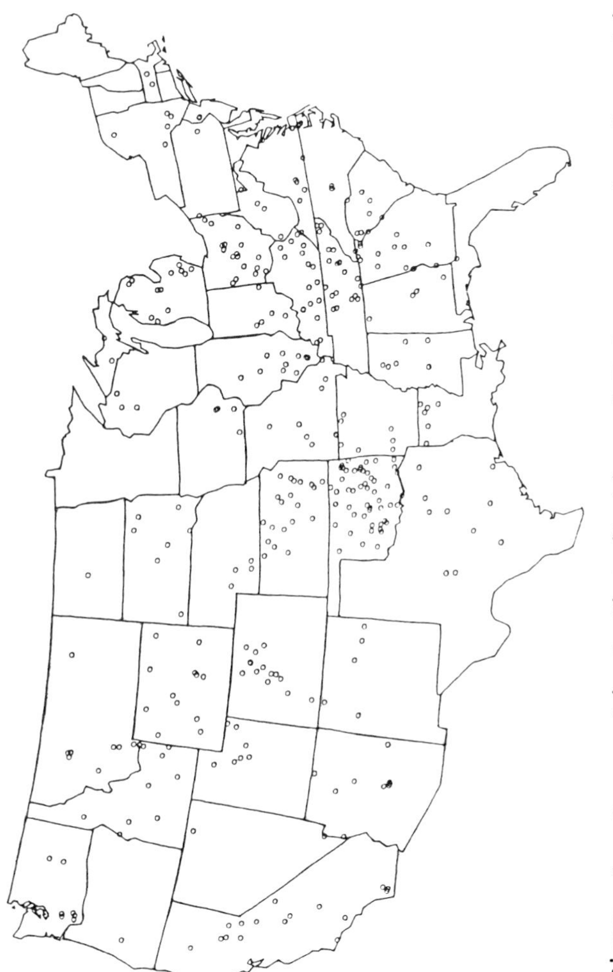

Figure 1. Geographical distribution of U.S. reservoirs from which angler use and sport fish yield data were obtained.

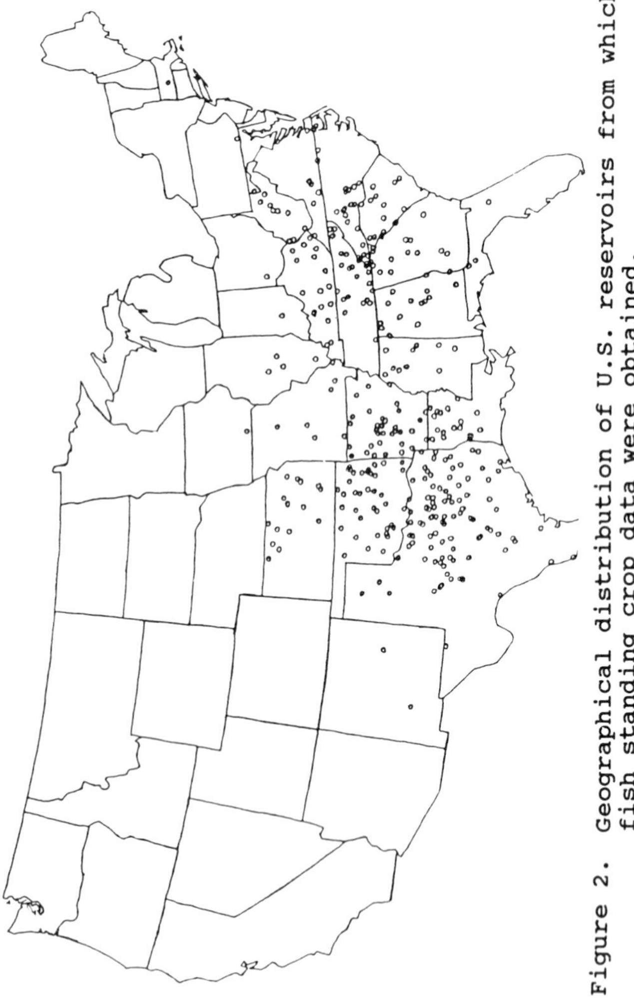

Figure 2. Geographical distribution of U.S. reservoirs from which fish standing crop data were obtained.

morphometry or fertility of a reservoir. Variables such as shore development, area, mean depth, outlet depth, and water-level fluctuation were mean annual constants for each reservoir. Storage ratios (volume/total annual discharge of water) and concentrations of total dissolved solids were calculated from USGS surface water records. Select water quality data were obtained from reports of the National Eutrophication Survey (U.S. Environmental Protection Agency 1975, 1978) on 172 reservoirs for which fish standing crop or yield data were available. Samples were taken at fixed sites, depths and seasons for one year and include measurements of Secchi disk transparency, phosphorus and nitrogen loading, and concentrations of total phosphorus, total nitrogen, inorganic nitrogen, alkalinity and chlorophyll "a". Water-quality data also were obtained from USGS records of samples collected immediately above reservoirs in tributaries or in tail-waters of mainstream reservoirs with high rates of water exchange (i.e. more than six times per year). Concentrations of chemicals in tailwaters of most mainstream reservoirs did not differ significantly from those in tributaries.

Fishery data include:

1. angler use and sport fish yield by reservoir, year and species

2. fish standing crop by reservoir, year and species.

The angler use and sport fish yield data base contains 1,327 years of records from 380 impoundments in 46 states (Figure 1). They represent 22% of the number and 46% of the total area of United States reservoirs. Angler use statistics include hours or days of angling pressure per acre and year, and the length of an average angler day. Yield statistics include the total number of weight of fish harvested per hour or day and yield in pounds per acre by species or species group. Estimates of angler use and sport fish yield include a variety of creel survey designs, sampling intervals and methods of reporting.

The fish standing crop data base cover 393 reservoirs and contains 1,973 years of record, many of which are averages of two or more cove samples collected from a reservoir during the same month and year. About 23% of the number and 45% of the total reservoir area existing in the U.S. is represented. Most of the standing crop estimates are from warm-water species, because rotenone sampling of coves has been used most extensively in the southern United States (Figure 2). Summer water temperatures must exceed 25°C for efficient use of this sampling technique.

Fish standing crop data were obtained from samples of fish recovered following treatment with rotenone or other piscicides. Most samples were collected from 1955 to 1985 by state and Federal agencies. Most agencies used large, fine mesh block nets to prevent fish from entering or escaping from the sampling area during treatment. Data represent the mean weight of fishes recovered per acre of cove area sampled.

ANGLER USE, YIELD AND STANDING CROP PREDICTION

The most important aspect of contaminant effects to most U.S. fishery management agencies is the effect on sport fish catch, angler use, and on fish consumption by humans. Reservoirs figure prominently in total aquatic resource use (Table 1). Therefore, readily measured indexes to angler success and quality and the quantity of fishes harvested should be of value in identifying perturbed environments. To demonstrate the utility and applicability of regression models to a wide range of lake conditions, predictions were made for a natural lake and a number of reservoirs.

Leech Lake

Regression models were used to make predictions on this uncontaminated 45,124 ha natural lake in Minnesota, where data on fish yield and environmental attributes were readily available (Schupp 1978). As shown in Table 2, Schupp's field estimate of annual sport fish yield was 4.48 kg/ha. Our predicted yield was 4.75 kg/ha based on a formula with the following independent variables: surface area (-), mean depth (-), age (-), and growing season length (+), (N = 336, R^2 = 0.29). Schupp's estimate of angler-hours/ha has 17.9 compared to our predicted 23.8; catch-rate was 0.26 kg/hour, compared to our predicted 0.26 kg/hour. The yield of walleyes was 2.2 kg/ha, compared to a predicted 3.9 kg/ha derived from a formula with the following variables: total dissolved solids (+), storage ration (+), growing season length (-), and shore development (-). Area was not a variable, which may account for our overestimate, as Leech Lake was near the maximum size of reservoirs in the sample used to develop predictive equations. Predicted yield equalled 28% of

TABLE 1 SURFACE AREA OF US FRESHWATERS BY TYPE AND PERCENT OF TOTAL FRESHWATER ANGLING ON EACH IN 1980 (USFWS 1982).

	Millions of hectares	Percent of total freshwater angling (1980)
Natural lakes and ponds, less Great Lakes and Alaska	3.8	21%
Large reservoirs >200 ha	4.0	38%
Small reservoirs 4 to 202 ha	2.3	
Man-made ponds <4 ha	1.0	15%
Rivers and streams	2.4	20%
Great Lakes (U.S.)	15.7	6%
Total	29.2	710 million angler-days

TABLE 2 COMPARISON OF FIELD ESTIMATES OF YIELD FROM LEECH LAKE, MINNESOTA (Schupp 1978) WITH THAT PREDICTED FROM REGRESSION ANALYSES OF U.S. RESERVOIR YIELDS (Ploskey et al. in press)

	Leech Lake	
	Field estimate (Schupp 1978)	Predicted (Ploskey, et al., In press)
Total sport fish yield	4.48 kg/ha/yr	4.75 kg/ha/yr
Walleye yield	2.2 kg/ha/yr	3.9 kg/ha/yr
Angler-hours	17.9/ha	23.8/ha
Angler catch rate	0.25 kg/hr	0.26 kg/hr

a predicted walleye standing crop of 14 kg/ha, derived from a formula with four independent variables: storage ration (+), outlet depth (+), growing season (-), and shore development (-). The yield/crop proportion appears quite reasonable.

Beaver Lake

The applicability of the yield prediction technique is illustrated by a comparison of the first 13 years of sport fish harvest records on a 11,410 ha reservoir in Arkansas, with predictions derived from a formula based on data from 46 reservoirs <28,330 ha, and a growing season >140 days (Figure 3). Although there is considerable scattering of observed yields from year-to-year, the 13 year mean is near the regression prediction. After the first 10 years of impoundment, major effects of contaminants might be detected by comparing sport fish yields to the mean predicted yield.

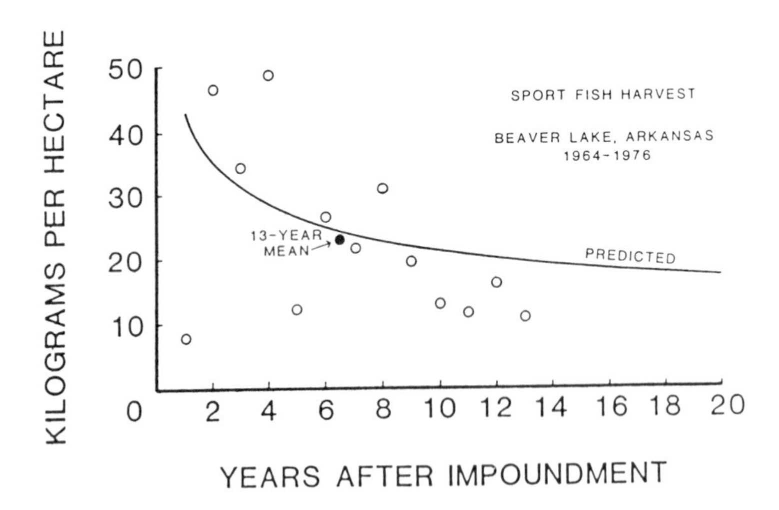

Figure 3. Field estimates of sport fish harvest, Beaver Lake, Arkansa, through 13 years of impoundment, compared with predicted harvest (from Jenkins 1977).

Belews Lake

The use of biomass predictors is illustrated by the effect of a contaminant on a private utility company reservoir, Belews Lake, a 1,564 ha cooling pond for a coal-fired plant in North Carolina which was monitored (physical, chemical, biological) following impoundment in 1971 (Olmsted et al. 1986). Samples of total fish crop from 1972 to 1974 averaged 98 kg/ha, compared to a predicted 114 kg/ha derived from the morphoedaphic index (Figure 4). In 1976, North Carolina Wildlife Resources Commission and Duke Power Company biologists noted a substantial decline in the abundance of adult fish and near absence of young-of-the-year fish in rotenone samples. Fly ash from the plant was routinely wet sluiced to a settling basin, and ash basin overflow into the reservoir began in 1975. Subsequent rotenone samples revealed a drastic decline in standing stock, averaging only 28 kg/ha from 1976 to 1981. Detailed toxicological studies identified high concentrations of selenium as the culprit and remedial measures were taken to prevent ash sluice waters from entering the reservoir. Continued monitoring should document the rate of recovery of the fish crops to predicted unstressed levels.

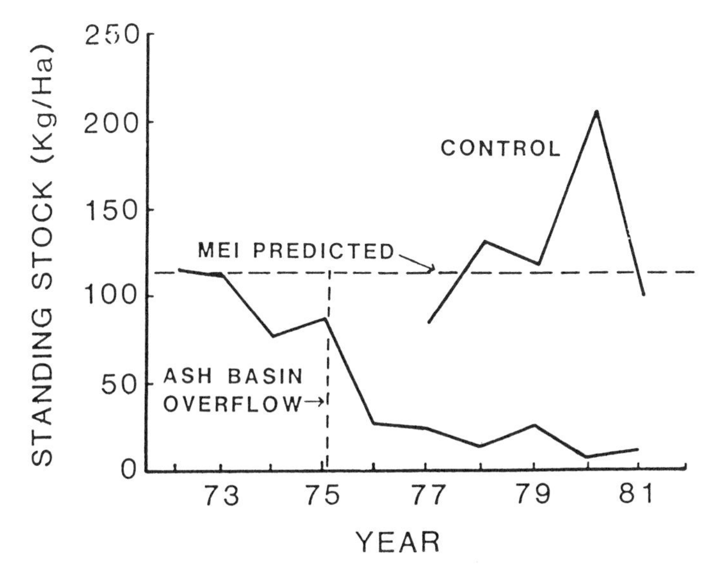

Figure 4. Standing crop of fish in Belews Lake, North Carolina, main lake and control coves, 1972-81, compared to mean crop predicted by the morphoedaphic index (adapted from Olmstead et al. 1986).

Lake Keowee

An example of using indexes to average biotic conditions to evaluate effects of heated effluents is available from Lake Keowee, a 7,400 ha cooling reservoir for a nuclear power plant in South Carolina, which was sampled annually with rotenone following impoundment. Total crop estimates from the second through the ninth year of impoundment showed a general decline with age (Figure 5). In the ninth year, the field estimate of fish crop equalled the predicted crop of 50 kg/ha. Without accumulated long-term data on typical trends in new reservoirs or a predicted crop based on the morphoedaphic or other indexes, the decline in fish crop might have been attributed to the effects of nuclear plant operations rather than to the aging of the reservoir.

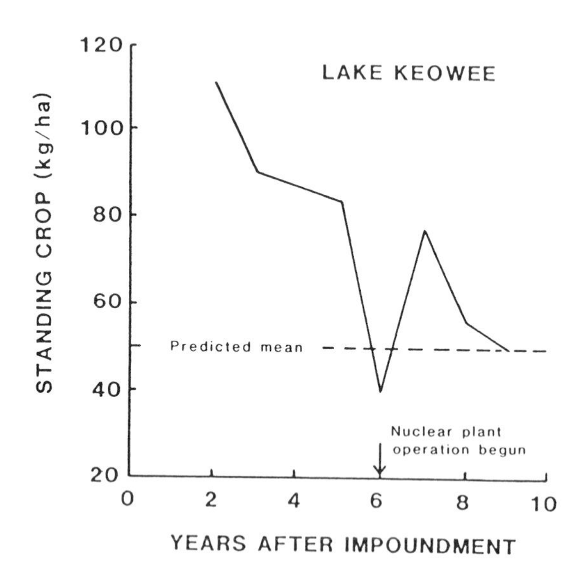

Figure 5. Field estimates of total standing crop in Lake Keowee, South Carolina, compared with the mean crop predicted by the morphoedaphic index (from Jenkins 1982).

Bull Shoals Lake

The largest known continuous standing crop data set for large impoundments is available on Bull Shoals Lake, an 18,400 ha reservoir in Arkansas-Missouri, which has been sampled annually since 1952. Based on three-year moving averages (Figure 6), estimates of total fish standing crop during the first 26 years of impoundment varied from 170 to 335 kg/ha, with a mean of 225 kg/ha. The morphoedaphic index value for the reservoir predicts a crop of 230 kg/ha (Jenkins 1982). Major departures from the mean were associated with high inflows and nutrient loadings in the 7th year of impoundment from an upstream 17,000 ha reservoir (Table Rock Lake) which filled before dam completion and overflowed into Bull Shoals Lake, and with extremely high water levels in the 16th and 23rd years.

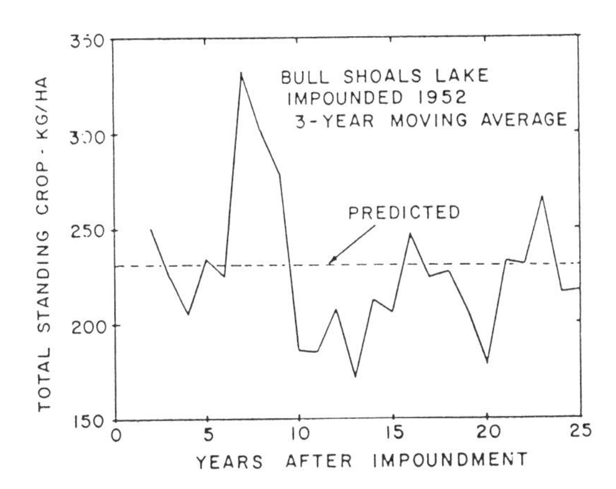

Figure 6. Total fish standing crop annual estimates, Bull Shoals Lake, Arkansas, 1952-77, compared to crop predicted by the morphoedaphic index (from Jenkins 1982).

WATER QUALITY INDICES

The national reservoir physiochemical data base (Ploskey et al. in press) was merged with water quality data base available from EPA's 1974-1975 National Eutrophication Survey, and regressed on yield and crop data from 112 reservoirs. Although the correlations do not denote direct cause-effect relationships, they are consistent and useful for predictive purposes. Some examples of variable relations in regressions: Concerning angler use and yield, total phosphorus content was positively related to the weight of fish harvested/hour and to yields of catfish and white bass. Secchi disk transparency was positively related to total yield, white bass, sunfish and black bass yields, and negatively to catfish yield. Chlorophyll "a" was positively related to total angler effort and yield and to carp, bullhead, striped bass, sunfish and largemouth bass yields. Alkalinity was positively related to catfish, white bass and walleye yields, and negatively to catch rate and yields of sunfish and black basses.

Concerning fish standing crop and water quality, negative relations included chlorophyll "a" versus walleye crop; Secchi disk transparency versus gars, carp, white crappie and freshwater drum crops; and alkalinity versus sucker and black crappie crops. Total clupeid, and sport fish crops were all positively related to total phosphorus, Secchi disk transparency, chlorophyll "a", and alkalinity. Many of these relationships are undoubtedly curvilinear, but have not been tested as yet. The development of additional formulae could help biologists describe optimum and detrimental levels of nutrients and turbidity in large natural lakes and separate their effects from toxic stress symptoms.

ALTERNATIVE APPROACHES

Carline (1986) reviewed the use of indices as predictors of fish standing crops, yields, and community structure and summarized existing simple models which may account for up to 90% of the variability in yield or standing crop from a subset of lakes. He urged exploration of new ways to describe fish community structure besides taxonomic groupings: e.g. trophic level, habitat use and reproductive mode. In proposing the application of biological indicators of ecosystem quality to the Great Lakes, Ryder and Edwards (1985) noted that trophic structure and particle-size density are two promising avenues for assessment of fish community status. Both differ substantially from the conventional species-by-species approach. They recommended lake trout and walleye as biological indicator and integrator organisms, surrogates of their reproductive communities adapted to oligotrophic and mesotrophic environments, respectively. As an alternative, the authors noted that opportunities exist for both creativity and rapid advancement in the direction of ecosystem assessment utilizing whole system properties and recommended:

> ..."the further development of this approach through detailed analyses of both quantitative and qualitative changes in Great Lakes fish yields during the last century and relating these to the concomitant environmental changes."...

Any of the above approaches could be pursued in large lakes in attempting to identify contaminant effects. As the late F.H. Rigler admonished (1982):

> ..."We acknowledge the challenge of the empirical approach and recognize its essential role in exposing predictable regularities in natural systems and in stimulating further empirical and explanatory studies...The only predictions we can collectively make for society derive from empirical theory...With a proper appreciation of the significance of empiricism in environmental science we will...accelerate the development of new predictive theories that are so essential to any society that intends to base its environmental decisions on predictions of science rather than whim."...

CONCLUSION

In evaluating ecological approaches to stressed fisheries, Paloheimo and Regier (1982) concluded that:

> ..."findings from our empirical, comparative modelling approach...demonstrate the overriding importance of the morphological and limnological variables in determining the structure of fish and planktonic communities. The effects of man-made stresses on the biotic community often seems to be mediated through their effects on abiotic factors...Any attempts to explain changes in multispecies fisheries that ignore concomitant changes in abiotic variables appear to be futile."...

The predictive utility afforded by relatively simple empirical models should be adequate for a number of practical purposes where budgetary constraints preclude detailed studies of a multitude of lakes. Long-term monitoring of fish populations will be required to identify changes which can be positively attributed to chronic toxicant effects. Economically feasible monitoring in large waters necessitates selection of a comparatively small number of variables and adherence to statistically sound procedures that will yield testable results. Ongoing monitoring of the physiochemistry and fish yields of most large natural lakes in the world could provide a data base for regression analysis and the development of indexes to lake ecosystem well-being. This has been done for North American glacial lakes by Ryder (1965), large reservoirs in Africa (Henderson and Welcomme 1974), and is being expanded for tropical and subtropical waters globally by FAO/United Nations, Rome (James M. Kapetsky, personal communication). The author recommends the development of such indexes of fish stock and yield until accelerated research can greatly refine our ability to detect the chronic effects of contaminants on large lake biota.

REFERENCES

Brown, J.G. et al. 1964. World register of dams, International Commission on Large Dams. Paris, 1964.

Bue, C.D. 1963. Principal lakes of the United States. Geological Survey Circular 476. U.S. Geological Survey, Washington, D.C. Fifth printing, 1970.

Carline, R.F. 1986. Indices as predictors of fish community traits. pp. 46-56 In: G.E.Hall and M.J. Van Den Avyle, editors. Reservoir fisheries management: strategies for the '80s. Reservoir Committee, Southern Division American Fisheries Society, Bethesda, Maryland, USA.

Fels, E. and R. Kellar 1973. World register on man-made lakes, In: Man-made lakes: Their problems and environmental effects. Geophysical Monograph Series, Vol. 17, edited by W.C. Ackermann, G.E. White and E.B. Worthington, pp. 43-49, AGU, Washington,D.C.

Gilbertson, M. 1984. Need for development of epidemiology for chemically induced diseases in fish in Canada. Canadian Journal of Fisheries and Aquatic Sciences 41:1534-1439.

Henderson, H.F. and R.L. Welcomme 1974. The relationship of yield to morphoedaphic index and numbers of fishermen in Africa inland fisheries. CIFA Occasional Paper No. 1, Food and Agricultural Organization of the United Nations, Rome.

Jenkins, R.M. 1977. Prediction of fish biomass, harvest and prey-predator relations in reservoirs. pp. 282 - 293 In: Webb Van Winkle (ed.) Proceedings of the conference on assessing the effects of power-plant induced mortality on fish populations. Pergamon Press, New York.

Jenkins, R.M. 1982. The morphoedaphic index and reservoir fish production. Transactions of the American Fisheries Society 111(2):133-140.

Martin, R.O.R. and R.L. Hanson 1966. Reservoirs in the United States. Geological Survey Water-Supply Paper 1938. U.S. Government Printing Office, Washington.

Olmsted, L.L., D.J. Degan, J.S. Carter and P.M. Cumbie 1986. Ash basin effluents as a concern of fisheries managers: a case history and perspective. pp. 261-269 In: G.E. Hall and M.J. Van Den Avyle, editors. Reservoir fisheries management: strategies for the 80's. Reservoir Committee, Southern Division American Fisheries Society, Bethesda, Maryland, USA.

Paloheimo, J.E. and H.A. Regier 1982. Ecological approaches to stressed multispecies fisheries resources. Canadian Special Publication of Fisheries and Aquatic Sciences 59:127-132.

Ploskey, G.R., L.R. Aggus, W.M. Bivin and R.M. Jenkins. In press. Regression equations for predicting fish standing crop, angler use and sport fish yield for United States reservoirs. Administrative Report, 86-__, Great Lakes Fishery Laboratory. U.S. Fish and Wildlife Service, Ann Arbor, Michigan, USA.

Rigler, F. H. 1982. The relation between fisheries management and limnology. Transactions of the American Fisheries Society 111(2):121-132.

Ryan, P.M. and H.H. Harvey 1977. Growth of the rock bass, *Ambloplites rupestris* in relation to the morphoedaphic index as an indicator of an environmental stress. Journal Fisheries Research Board of Canada 34:2079-2088.

Ryder, R.A. 1965. A method for estimating the potential fish production of north-temperate lakes. Transactions of the American Fisheries Society 94:214-218.

Ryder, R.A. 1982. The morphoedaphic index - use, abuse and fundamental concepts. Transactions of the American Fisheries Society 111:154-164.

Ryder, R.A., S.R. Kerr, K.H. Loftus and H.A. Regier 1974. The morphoedaphic index, a fish yield-estimate: review and evaluation. Journal of the Fisheries Research Board of Canada. 31:663-688.

Schupp, D.H. 1978. Walleye abundance, growth, movement and yield in disparate environments within a Minnesota lake. Pages 58-65 In: R.L. Kendall, editor. Selected coolwater fishes in North America. American Fisheries Society Special Publication No. 11.

U.S. Environmental Protection Agency. 1975-1978. Compendia of lake and reservoir data collected by the National Eutrophication Survey. Working papers 474-477. Corvallis, oregon.

USFWS (U.S. Department of the Interior, Fish and Wildlife Service and U.S. Department of Commerce, Bureau of the Census) 1982. 1980 national survey of fishing, hunting and wildlife-associated recreation. U.S. Government Printing Office, Washington, D.C. USA.

Welch, P.S. 1935. Limnology. Mc-Graw-Hill xiv + 471 pp.

Work Group on Indicators of Ecosystem Quality 1985. A conceptual approach for the application of biological indicators of ecosystem quality in the Great Lakes Basin. R.A. Ryder and C.J. Edwards, editors. Report of the Aquatic Ecosystem Objectives Committee to the Great Lakes Science Advisory Board of the International Joint Commission and the Great Lakes Fishery Commission. 169 pp.

CHAPTER 11

PERSPECTIVES ON THE INFLUENCE OF TOXIC SUBSTANCES ON FISHERY PRODUCTIVITY

George R. Spangler

Department of Fisheries and Wildlife, University of Minnesota, St. Paul, MN.

ABSTRACT

It is a fond illusion shared by many fishery biologists that the traditional exponential formulation for expressing mortality in fish populations might eventually be extended to include all factors thought to be responsible for significant mortality. This school of thought has resulted in both high and low resolution models useful for the simulation of yield reduction due to contaminant burdens. These models, based on traditional age-structured population theory, currently provide the most convenient estimates of toxic substance impairment of fishery yield. An analogue of the Schaefer surplus production model applied to salmonid prey fish abundance in the Great Lakes may provide a mechanism to detect a stock's latitude for compensation to toxic substances.

Repeated failure to achieve only moderately precise yield estimates, even for factors as important as sea lamprey (*Petromyzon marinus*) predation on Great Lakes fish, has demonstrated the need for development of more holistic models incorporating variables sensitive to population perpetuation. Very few holistic models have been applied to the problem of estimating contaminant effects, but recent progress in measuring latent mortality and sub-lethal contaminant effects suggest that it not inappropriate to infer major contaminant influences from minor effects exhibited in the laboratory. Particle-size distribution models and general yield models of the morphoedaphic index type may be useful in estimating losses of fishery yield due to contaminants.

In spite of the uncertainties associated with extrapolation of laboratory toxicity results to wild populations, simulations of anticipated impact have been applied to estimation of fishery losses in the past and will continue to provide low resolution estimates of short-term economic losses to the fishery where clear population compensatory processes can be identified.

INTRODUCTION

Environmental factors affecting fish stocks consist of all possible combinations of physical, chemical and biological entities acting concurrently or consecutively (Sindermann 1984). Among these factors are the "environmental contaminants" or anthropogenic compounds refined or synthesized specifically for biotic control purposes (fertilizers, pesticides) or as manufacturing process byproducts not intended to result in biological responses (e.g. polychlorinated biphenyls - PCBs, polybrominated biphenyls - PBBs). Also included are naturally occurring substances such as the heavy metals that have been rendered more available to the biosphere by way of human activity. Some factors are acutely lethal to individual fish whereas others may lead to debilitation, physiological malfunction or chronic effects that recur periodically such that the overall effect is a degradation of the productive capacity of the stock or population. This paper examines a number of explicit conceptual models that may be used to infer detrimental population effects from compounds known or expected to be deleterious to individual organisms.

Since the publication of **Silent Spring**, the "public" environmental awareness in North America has focused upon the negative effects of contaminants, and that is the central issue here, but the broader concern is to examine ways of understanding the effects of these compounds at levels of organization beyond individual organisms, irrespective of societal judgment as to their positive or negative value.

Individual organisms may be seen as the "emergent properties" of natural selection in Darwinian evolution; so too are they the basic units for expression of the effects of exposure to environmental contaminants. By analogy we might argue that just as species are the emergent properties of evolution, so must there be emergent properties of populations and communities of organisms (Kerr 1974). The challenge to modern ecologists is then to define these properties and to examine them for the "stress" reponses that are induced by environmental contaminants (Rapport 1984).

The problem of contaminant effect has historically been approached through the organismal definition of stress due to (Selye 1950, 1952):

> ..."the sum of all the physiological responses by which an animal tries to maintain or re-establish a normal metabolism in the face of a physical or chemical force."...

This has been extended by Esch et al. (1975) to "any level" of biological organization for which stress is the product, not the cause of homeostatic change, i.e.:

> ..."stress is the effect of any force which tends to extend any hemostatic or stabilizing process beyond its normal limit"...

Wedemeyer et al. (1984) and Colby (1984) elaborate this point further and provide functional definitions for stress, stressors and population compensation to stress. The current task is to evaluate what progress has been made in understanding population or community-level responses to environmental contaminants.

The need for tools to examine the stress responses of biological entities above the organismal level of organization (Kerr and Dickie 1984) cannot be overstated, but this is not to say that valuable inferences cannot be made within the context of the traditional organismal or population models. Our interest in the differences between levels of biological organization in their responses to environmental contaminants is two-fold:

1. most existing methods of measuring effects have been developed at the biochemical or physiological levels of organization, and this will probably continue to be the case in the near future, and

2. the duration of effects and the latency of their expression may be so protracted that enormous environmental damage is done before

we recognize the consequences of our misdeeds (see especially Sindermann 1984, Figure 3).

The latter factor lends an urgency to this task that demands an approach which can build upon existing knowledge while new conceptual models are being developed.

ESTIMATING CONTAMINANT EFFECTS

Two divergent schools of thought have been responsible for the models currently available in the fisheries and ecological literature. Both approaches have their merits and relative disadvantages, but each offers models of sufficient resolution to be useful in simulating the effects of environmental contaminants on fishery yield. Development of these models and innovative laboratory procedures for determination of sublethal effects will greatly improve our understanding of the impact of anthropogenic environmental contaminants.

Population Model Paradigm

Traditional models applied to fisheries are presented here as high or low resolution according to their requirements for input (the data stream used to estimate model parameters) information. Each of these types of models may provide the basis for simulation of the outcome of a number of different hypothetical contaminant inputs. Simulation models may also be constructed from combinations of model types where high resolution information may be reduced or aggregated sequentially within the model (generally in a discrete sub-model of the overall simulation exercise) in preparation for delivery to another model (or sub-model) type of lower resolution. This mixing of model types is not reversible, i.e. estimates will exist only for model parameters of a resolution equal to, or lower than, those possible from the highest resolution input information. These models have originated from consideration of yield estimation problems associated with individual species. Multi-species applications of these models have been little more than aggregations of a number of the single-species formulations with their separate parameter estimates. Biological linkages between species have been speculative at best, except for the bioeconomic relationships influencing the selective exploitation of stocks by commercial fisheries (Clark 1985). Vaughan et al. (1984) summarize the mathematical formulations for the fisheries models most commonly used in development of yield forecasting simulation models.

The highest resolution models available include the age and size-structured analytical models that segregate the major population processes of growth, mortality and reproduction into discrete functions, each with its own suite of parameters. The yield models of Beverton and Holt and Ricker that balance individual body growth and cohort mortality are examples of high-resolution models. Explicit consideration of the relationship between stock and recruitment and a life history stage-structuring (as in Leslie Matrix formulations) provide the flexibility necessary to simulate population changes that will occur in future generations given specific information relating the presence of contaminants to major population processes. These models typically represent known sources of mortality in the differential equation:

$$\frac{dN(t)}{dt} = -M * N(t)$$

where the rate of change in numbers in the population (Left side of equation) is a function of the population size N (t) multiplied by the known sources of mortality M. If M consists of a number of different factors, it is a simple matter to represent their aggregate effects as a sum:

$$M = \sum_i m_i$$

This formulation allows the mortality factor to be defined in any way appropriate to circumstances. For example, the effect of fishing is frequently expressed (after appropriate definition of the interval over which it influences the stock) as a "special" mortality coefficient F, but the individual m_i terms might just as easily represent mortality due to things like DDT or PCB burden. While there is no explicit provision in these models for effects such as genetic change in a stock, this contingency can be handled if the change can be related to the growth, mortality or reproduction functions. Latent mortality due to contaminant burdens, or slow growth due to physiological impairment simply modify the parameters of these functions in the overall simulation model.

Recently, major advances have been made in cytogenetic methods for the detection of mutagenic and carcinogenic effects of contaminants (Hollstein et al. 1979, Landolt and Kocan 1983). Hendricks et al. (1980) reported a latent (one year post-treatment) dose-dependent response in tumor formation in rainbow trout, *Salmo gairdneri*, exposed for only one hour to MNNG (M-methyl-N'Nitro-N'-Nitrosoguanidine) concentrations ranging from 10 to 100 ppm. Similar exposures yield anaphase aberrations detectable by **in-vivo** assays within 24 hours of treatment (R. Phillips, Dept of Biology, Univ. Wisconsin-Milwaukee, pers. comm.) Ability to detect the effects of genotoxic agents at dosages below those which cause significant (short-term) mortalities will enable the application of laboratory test results for these contaminants to simulations of population effects occurring at advanced ages. It is particularly important to note that these results were obtained at very low dosages of short duration, conditions approaching those of chronic contaminant exposure for many Great Lakes fishes in nearshore waters. Clearly, short-term exposure events such as point source contamination from even brief releases of industrial effluents, can have significant consequences for the health of fishes in the vicinity of release points. Events of this kind can be input to the population simulation models through modification of the mortality coefficients for the juvenile life history stages, or, if the contaminants exhibit biological impairment of gamete viability, their effects can be modelled through the reproduction functions in the simulation models.

Examples of interesting results from high-resolution population models are provided by the Chesapeake Bay striped bass (*Morone saxatilus*) simulation (Goodyear 1985) and the lake trout (*Salvelinus namaycush*) rehabilitation model (Walters 1986). Westin et al. (1985) report empirical estimates of striped bass larval mortality associated with body burdens of four organochlorine contaminants. These are the types of estimates that could be input to Goodyear's model which uses larval fish mortality corrections to represent impacts due to contaminant burdens.

Walters' simulation model (Source coding in Applesoft™ for Apple II microcomputers available from the Great Lakes Fishery Commission, 1451 Green Road, Ann Arbor,MI 48105) is particularly appropriate for examining the long-term population effects of contaminants. The model was developed to depict the time-course for rehabilitation of the lake trout stocks for the upper Great Lakes, given some of the uncertainties surrounding the reproductive viability of hatchery fish release in the lakes. The model includes explicit functions for growth and reproduction, and tables for fecundity and age-specific mortality. Reproductive impairment is represented by assuming that "wild" fish are 100 percent effective as spawners whereas planted fish will be somewhat less effective, e.g. 50% or whatever the modeller wishes to explore with the simulation. In addition to the reproductive equations, the number of juvenile trout is a function of the growth of females to the size and age of first spawning. Thus any environmental factor reducing growth or increasing the mortality of the females will diminish the reproductive contribution of that year-class of fish (stocked or wild). The "normal" growth function assumed in the model yields growth rates intermediate between the extremes of fast and slow growth for lake trout in Lake Superior from 1948 to the early 1960s.

The model produces 30-year simulations for a stock of lake trout consisting initially of planted fish only. Fisheries managers are interested in estimates of the numbers of fish that will be taken by the fishery and knowing the progress of stock rehabilitation, i.e. rate of approach to self-sustaining stocks, as a function of fishery regulation, lamprey control and numbers of fish planted annually. Fixing all factors in the model except for growth and reproductive impairment produced the simulations presented in Figure 1. Four scenarios were examined depicting the combinations of normal and slow growth (ng, sg, respectively) and moderate (20%) reproductive impairment versus none.

Two results are immediately apparent. First, the rehabilitation process for such slow-maturing fish as lake trout will not yield a significant harvest of wild fish for at least 15 years even under the most optimistic simulation conditions. Secondly, the effect of slower growth and the lowering of fertility exacerbate the recovery process to an extent that precludes any reasonable expectation of ever achieving a rehabilitated stock (less than 30% of the yearlings would be wild fish even at normal rates of growth). Note that it is not necessary to stipulate the nature of the reproductive impairment, only the magnitude is important. Thus, contaminant burdens that reduce egg hatchability (Macek 1968) or influence behaviour such as temperature preference (Mac and Bergstedt 1981) will be reflected in delays in the progress of rehabilitation.

At the very least, simulations for economic impact of reduced fish production due to contaminants can usually be carried out. Van Oosten (1928) estimated the economic loss to the Saginaw Bay shallow-water cisco

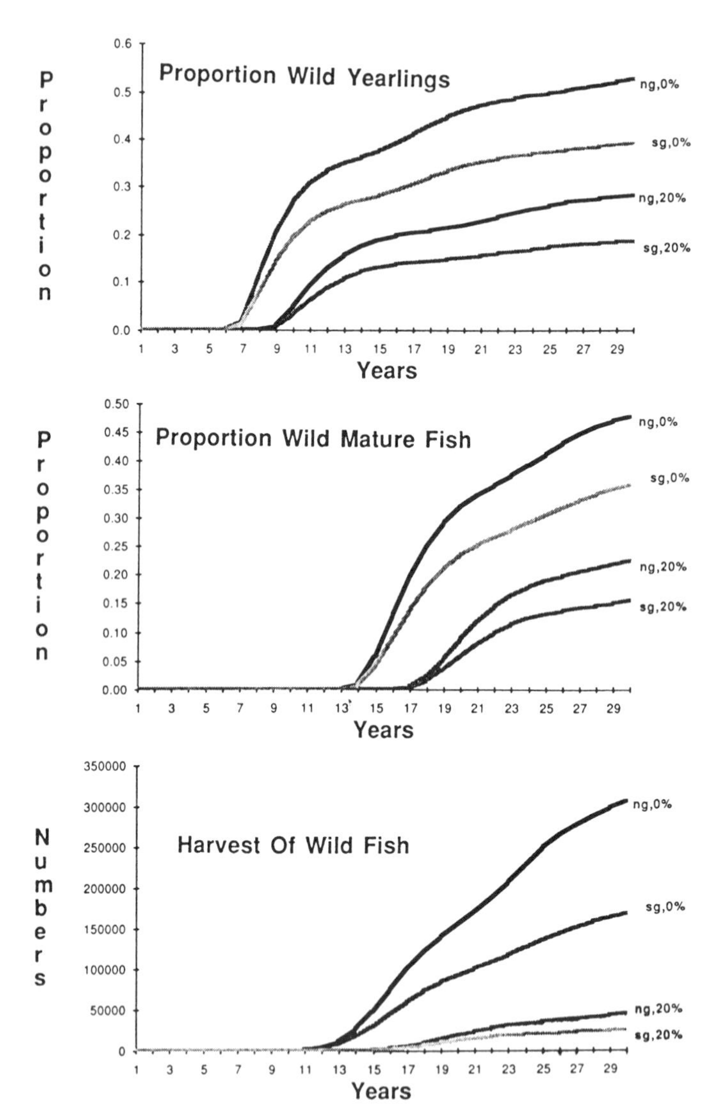

Figure 1. Simulated progress toward a rehabilitated lake trout community as revealed by wild fish in the catch and in the population of yearlings and mature fish. Slow growth (sg), normal growth (ng) and reproductive impairment (0% and 20% reduction in viability) as a result of contaminant burdens are simulated in 30 year scenarios.

(_Coregonus artedi_) fishery by simply knowing typical growth and mortality schedules for the stock before and after the contaminant spill. This approach is just as valid today and much simpler to accomplish with modern computing equipment available to every biologist.

An alternative to the detailed population dynamics models underlying simulations of the lake trout and striped bass populations are the so-called "surplus production" models based on the notion that exploited fish stocks yield biomass to a fishery in proportion to the difference between stock size and the "carrying capacity" of the area that supports the stock (Schaefer 1954). These models are low-resolution formulations that inextricably confound the growth, reproduction and mortality functions of the exploited stocks. This is an attractive feature from the standpoint of parameter estimation, essentially requiring only catch and effort information from the fishery. It is also interesting from an environmental perspective, because such models would seem to directly address the question of how much yield (surplus production) would be lost due to contaminant burdens. This is a mixed blessing in the latter application since there is no clear way to relate laboratory information on acute or chronic toxicity to the basic model parameters of carrying capacity or intrinsic growth rate. On the other hand, behavioral studies that demonstrate an influence of contaminants on a fish's inherent vulnerability to the fishery could be input directly to these models through their "catchability coefficient" parameter. Most direct of all would be an extensive time series of catch and effort data over different time intervals in which contaminants are known to have been present in one case and absent in the other. Contrasting the apparent change in surplus production between these periods would then provide some insight into the question of how the contaminants may have altered the carrying capacity of the environment.

Surplus production models have not been successful in describing the dynamics of Great Lakes fisheries. The generally accepted explanation (Walter 1986) is that the fish stocks do not come into equilibrium with the fishery. This is certainly the case for fisheries that depend upon the top level consumers in the aquatic community, e.g. whitefish, _Coregonus clupeaformis_ and lake trout. This is probably not the case for species that are lower in the food chain. Certainly zooplankton and phytoplankton biomasses exhibit some degree of "stability" from one lake to another, hence the apparent differences in overall productivity between the lakes are maintained between years. The key to application of the surplus yield models would seem to be in choosing organisms or groups of organisms with life histories that are short relative to the time it takes the fisheries to respond to changes in population abundance. This dictum would apply also to surplus "yield" formulations of the relationship between piscivorous species and their prey. Thus, it seems reasonable to examine the relationship between salmonid abundance, known from planting records in the upper Great Lakes and the "surplus production" of their prey. The principle species of prey fish, alewives (*Alosa pseudoharengus*), smelt (*Osmerus mordax*), and the deepwater coregonines of Lake Michigan might show different estimates of surplus production than would the smelt and coregonines from Lake Superior. Differing levels of organochlorine contaminant burdens in the various Great Lakes may explain to some extent whatever differences are found in prey fish production.

Models offering varying degrees of resolution between the extremes of the age-structured population dynamics type and the simpler surplus production models are also available. In particular, cohort analysis (Pope 1972), the Deriso delay-difference model (Deriso 1980) and the general model of Schnute (1985), which subsumes all of the formulations discussed above, provide forecasting possibilities for simulating stock and population responses to contaminant inputs.

Trophic Dynamics Paradigm

Alternatives to the species typology inherent in the population approach are the various models relating to energy transfer through aquatic ecosystems. Three major classes of models are available, again ranging from highest resolution to lowest:

1. particle-size spectrum models,
2. bio-energetic models, and
3. morpho-edaphic production indices.

These models should be especially useful for application to systems for which the taxonomy of the organisms is unclear, e.g. benthic invertebrate studies involving chironomids or Great Lakes fisheries involving the deep-water Coregonines.

Particle-size spectrum applications are still in their infancy despite significant advances in theory (Kerr and Dickie 1984). Scientific interest in the particle size spectrum stems from observations of the marine pelagic community by Sheldon et al. (1972) who found a slight linear biomass decrease with increasing body size among successive groups of size-classes ranging from bacteria to whales. The basis for the method's application to environmental contaminant effects would be first to determine what are the appropriate or "characteristic" size spectra for healthy aquatic communities, and then to examine departures from that standard as a function of the presence of contaminants (Borgmann 1985). Data required for these models are field samples drawn from all available size-classes of organisms in the aquatic community.

The bio-energetic approach can model changes on the basis of size as well as age. These models typically decompose the change in individual biomass into the physiological processes of consumption, respiration, excretion, egestion and gamete production. These models offer the clearest resolution of the flow of contaminants through the ecosystem into top-level piscivores (Weininger 1978). Input information may be developed for individual species measurements as in the sea lamprey (Petromyzon marinus) application of Kitchell and Breck (1980) or it may derive from generally accepted rules such as the Q-10 respiratory coefficient for factors which are well-understood for a variety of different organisms. The greatest advantage of these models for assessing contaminant impacts is that sub-lethal effects manifested in observable physiological responses of laboratory animals may be incorporated into model predictions that translate into population effects through the growth or reproduction functions. We might also speculate that such constructs would be a useful modelling framework for community level effects when (if) we are able to develop reliable indices of community metabolism.

The tropho-dynamic approach includes the lake typology models such as the morphoedaphic index (MEI) useful in estimating fishery productivity on the basis of an individual lake's morphological and edaphic characteristics (Ryder 1965, 1982). Application of these models to contaminant effects problems would proceed as in the case of particle size-spectrum models, i.e. determining what level of fishery productivity is appropriate for a given lake and contrasting this with realized yield as a function of contaminant burdens. The advantage of this approach is the direct estimation of the parameter of interest, i.e. loss of yield due to contaminants. The principal disadvantage is the characteristically moderate precision of yield estimates based on these indices and the inevitable confounding of non-contaminant effects with other sources of stress on the fishery.

CONCLUSIONS

The influence of fish production of anthropogenic contaminants in the aquatic environment has not been directly estimable because the biological consequences of exposure have been measured at the physiological or organismal levels of organization. Simulation studies to develop useful estimates for higher levels of organization can be based upon either the traditional population dynamics paradigm with both high and low resolution models, or upon the trophic dynamics paradigm. Both of these approaches require additional research to define the cause and effect relationship between contaminant burdens and fish health. Additional exploration of the more holistic surplus production models may yield direct estimates of the extent to which fish yields are being compromised by current levels of contaminants in the Great Lakes.

ACKNOWLEDGEMENTS

I am indebted to Mr. Larry Jacobson who generously conducted the tedious simulations with the lake trout rehabilitation model. This work was supported by the Minnesota Agricultural Experiment Station.

REFERENCES

Borgmann, U. 1985. Predicting the effect of toxic substances on pelagic ecosystems. Sci. Total Environ. 44;111-121.

Clark, C.W. 1985. Bioeconomic modelling and fisheries management. John Wiley & Sons, Inc., New York: 291 pp.

Colby, P.J. 1984. Appraising the status of fisheries. p. 233-257, In: Cairns, V.W., P.V. Hodson and J.O. Nriagu (eds.) Contaminant effects on fisheries. John Wiley & Sons, Inc. New York. 333pp.

Deriso, R.B. 1980. Harvesting strategies and parameter estimation for an age-structured model. Can. J. Fish. Aquat. Sci. 37:268-282.

Esch, G.W., J.W. Gibbons and J.E. Bourque 1975. An analysis of the relationship between stress and parasites. Am. Midl. Nat. 93:339-353.

Goodyear, C.P. 1985. Toxic materials, fishing and environmental variation: simulated effects on striped bass population trends. Trans. American. Fish. Soc. 114:107-113.

Hendricks, J.D., R.A. Scanlan, J.L. Williams, R.O. Sinnhuber and M.P. Grieco 1980. Carcinogenicity of N-methyl-N'-nitro-N-nitrosoguanidine to the livers and kidneys of rainbow trout (Salmo gairdneri) exposed as embryos. J. Natl. Cancer Inst. 64:1511-1515.

Hollstein, M., J. McCann, F.A. Angelosanto and W.W. Nichols 1979. Short-term tests for carcinogens and mutagens. Mutation Res. 65:133-226.

Kerr, S.R. 1974. Structural analysis for aquatic communities. Proc. First Int. Congress Ecol. Centre for Agricultural Publishing, Wageningen: 69-74.

Kerr, S.R. and L.M. Dickie 1984. Measuring the health of aquatic ecosystems. pp. 279-284, In: Cairns, V.W., P.V. Hodson, and J. O. Nriagu (eds.) Contaminant effects on fisheries. John Wiley & Sons, Inc. New York. 333 pp.

Kitchell, J.F. and J.E. Breck 1980. Bioenergenetics model and foraging hypothesis for sea lamprey (Petromyzon marinus). Can. J. Fish. Aquat. Sci. 37:2159-2168.

Landolt, M.L. and R.M. Kocan 1983. Fish cell cytogenetics: A measure of the genotoxic effects of environmental pollutants, pp. 335-353 In: Niagru, J.O. (ed.) Aquatic Toxicology. Adv. Env. Sci. and Tech., vol. 13. J. Wiley and sons, New York, N.Y.

Mac, M.J. and R.A. Bergstedt 1981. Temperature selection by young lake trout after chronic exposure to PCBs and DDE. Pages 33-35 In: Chlorinated hydrocarbons as a factor in the reproduction and survival of lake trout (Salvelinus namaycush) in Lake Michigan. U.S. Dept. Int. Fish and Wildl. Sev. Tech. Pap. 105: 42pp.

Macek, K.J. 1968. Reproduction in brook trout (Salvelinus fontinalis) fed sublethal concentrations of DDT. J. Fish. Res. Board Can. 25:1787-1796.

Pope, J.G. 1972. An investigation of the accuracy of virtual population analysis using cohort analysis. Bull. I-ICNAF, No. 9:65-74.

Rapport, D.J. 1984. State of ecosystem medicine, p. 315-324 In: Cairns, V.W., P.V. Hodson, and J.O. Nriagu (eds.) Contaminant effects on fisheries. John Wiley and Sons, Inc. New York. 333 pp.

Ryder, R.A. 1965. A method for estimating the potential fish production of north temperate lakes. Trans. American. Fish. Soc. 94:214-218.

Ryder, R.A. 1982. The morphoedaphic index--use, abuse and fundamental concepts. Trans. American Fish. Soc. 111:154-164.

Schaefer, M.B. 1954. Some aspects of the dynamics of populations important to the management of the commercial marine fisheries. Bull. Inter-Am. Trop. Tuna Comm. 1:25-56.

Schnute, J. 1985. A general theory for analysis of catch and effort data. Can. J. Fish. Aquat. Sci. 42:414-429.

Selye, H. 1950. Stress and the general adaptation syndrome. Br. Med. J. 1:1383-1392.

Selye, H. 1952. The story of the adaptation syndrome. Acta, Inc. Montreal, Quebec. 255p.

Sheldon, R.W., A.Prakash and W.H. Sutcliffe, Jr. 1972. The size distribution of particles in the ocean. Limnol Ocean. 17:327-340.

Sindermann, C.J. 1984. Fish and environmental impacts. Arch. Fisch Wiss. 35:125-160.

Van Oosten, J. 1928. Life history of the lake herring (Leuchichthys artedi Le Sueur) of Lake Huron as revealed by its scales, with a critique of the scale method. Bull. U.S. Bur. Fish. Vol. XLIV.

Vaughan, D.S., R.M.Yoshiyama, J.E. Breck and D.L. DeAngelis 1984. Modelling approaches for assessing the effects of stress on fish populations. pp 259-278 In: Cairns, V.W., P.V. Hodson and J.O. Nriagu (eds.), Contaminant effects on fisheries. John Wiley & Sons, Inc. New York 333 pp.

Walter, G.G. 1986. A robust approach to equilibrium yield curves. Can. J. Fish. Aquat. Sci. 43:1332-1339.

Walters, C.J. 1986. The lake trout rehabilitation model: program documentation. L.D. Jacobson and G.R. Spangler, (eds.) Spec. Pub. 86-1. Great Lakes Fishery Commission, Ann Arbor, MI.

Wedemeyer, G.A., D.J.McLeay and C.P. Goodyear 1984. Assessing the tolerance of fish and fish populations to environmental stress: the problems and methods of monitoring. pp. 163-195. In: Cairns, V.W., P.V. Hodson and J.O. Nriagu (eds.) Contaminant effects on fisheries. John Wiley & Sons, Inc. New York. 333 pp.

Weininger, D. 1978. Accumulation of PCBs by lake trout in Lake Michigan. Ph.D. Thesis (Water Chemistry). Univ. Wisconsin, Madison, WI. 232p.

Westin, D.T., C.E. Olney and B.A. Rogers 1985. Effects of parental and dietary organochlorines on survival and body burdens of striped bass larvae. Trans. American Fish. Soc. 114:125-136.

CHAPTER 12

RESTOCKING OF GREAT LAKES FISHES AND REDUCTION OF ENVIRONMENTAL CONTAMINANTS 1960-1980

Howard A. Tanner

MSU, East Lansing, Michigan, U.S.A.

ABSTRACT

The Great Lakes of North America total 260,000 square kilometers in area and constitutes a significant portion of the world's freshwater. From 1800 to 1950 this watershed was changed from a wilderness into the industrial heartland of North America. During this period there occurred extensive overexploitation of fish stocks, increased sedimentation and widespread pollution from cities and industry. Since 1950 pollution control programs have achieved major improvements in water quality. Contaminants with toxicity of long duration, DDT, PCB, mercury, dieldrin, toxaphene, mirex and others, have been the slowest to respond to cleanup efforts. The stocking of salmonids to restore the predacious elements of the food web, lost to excessive commercial fishing and the parasitic depredations of the sea lamprey, began in the late 1950s. The riparian states, Ontario and the federal governments of Canada and the U.S. have stocked chinook, coho, pink salmon, lake trout, steelhead, brown trout and Atlantic salmon. Because of contaminants the sale of fish has at times been restricted or banned. Sport anglers have received health warnings. The fisheries of the Great Lakes have been restored and expanded by restocking, reduced commercial fishing and sea lamprey control. Sportfishing generates annually at least $2 billion dollars in economic activity. This program, beset by the negative impacts of contamination, has included prudent restrictions and warnings to consumers consistent with levels of risk. Contaminants continue to decline and the health and economic worth of the fishery continue to improve.

INTRODUCTION

The five Great Lakes of North America contain an estimated twenty percent of the world's supply of surface freshwater. Shared by Canada and the United States, these lakes and their connecting waterways and watersheds are integral parts of what has become the industrial heartland of these two nations. This industrial complex, important agricultural production and urban centers have polluted and contaminated portions of lakes with a variety of wastes.

Historically the abundant fishery resources of these lakes have been badly abused by over-fishing, pollution, accidental and deliberate introduction of exotic fish species, the damming of tributary streams and the draining of marshes. Several important native fish species have become extinct; several more are listed as endangered; and the former distribution of others has been reduced. By the late 1950s these abuses had combined to produce a near collapse of most species and their associated fisheries.

Beginning in the early 1960s, programs to enhance the fisheries of these lakes have been carried out by the federal governments, the province of Ontario and the eight states riparian to the lakes. Today, almost 30 years later, the ecological balance between predator and prey species has been restored and is sustained by the annual stocking of many millions of salmon and trout. Sea lamprey (<u>Petromyzon marinus</u>) control programs are effective.

The management of the fisheries has in most areas been shifted away from commercial fishing to recreational or sport fishing. Presently the economic activity generated by this sport fishery is estimated at 2 billion dollars annually.

Throughout the development of this enormously important fishery, the problem of contamination by a variety of persistent waste products has been of concern. One after the other, DDT, PCB, mercury, dieldrin, toxaphene and others have been recognized and actions taken to eliminate or reduce the problem-causing materials.

Much progress has been made. Some problems remain. Throughout this period people have continued to consume fish from the Great Lakes, guided by a variety of bans, advisories and, in the case of fish for sale, by action levels. Contamination levels continue to decline but will remain a problem to be dealt with for many years to come. Airborne contaminants represent the largest remaining source of most pollution.

The purpose of this paper is to put forward a description of how stocking of selected species, shifts in regulations and rigorous cleanups of contamination have been employed to restore the fishery potential of these lakes to where they yield enormous economic and recreational values. A review of the corrective measures of the past 35 years is a useful assessment of what has been achieved and should provide useful guidance as we face the problems and opportunities of the future.

DESCRIPTION AND HISTORY

The watershed of the Great Lakes is about 780,000 square kilometers in area. The lakes with their connecting waters total almost 260,000 square kilometers. The soils of the watershed in the northwest tend to be shallow and poor in nutrients; the product of the granitic shield of Canada. The human population around the shores of Lake Superior and the northern reaches of Lake Michigan and Huron are sparse and the land largely remains forested. The heavier soils rich in nutrients and crop producing capability begin in southern Wisconsin, stretch all along the southern shores of Lake Michigan, the southern one-third of Michigan to Saginaw Bay, and the southern shores of Lake Huron. Ontario's rich soils and extensive agriculture begin on the southern shores of Lake Huron. These rich soils extend eastward around the shallow enriched water of Lake Erie and embrace much of Lake Ontario.

It is in the areas of these productive soils that most of the population of the watershed has chosen to live, to farm, to build cities and industry. As a consequence, this area contributes the most in volume of wastes and in diversity of contaminants.

The five lakes are connected by rivers. The St. Mary's River flows out of Lake Superior to Lake Huron. Lake Michigan is separated from Lake Huron only by the straits of Mackinac. Lake Michigan is essentially a cul-de-sac receiving inflow from a very limited watershed. The St. Clair River flows from the southern end of Lake Huron to Lake St. Clair, hence down the Detroit River to Lake Erie. At the eastern end of Lake Erie the Niagara River flows over Niagara Falls to Lake Ontario. The St. Lawrence River flows from the northeastern end of Lake Ontario to enter the tide waters of the Gulf of the St. Lawrence.

In the early 1950s, Canada and the United States were faced with the disastrous collapse of many Great Lakes fish stocks. Response included the establishment by treaty of the Great Lakes Fishery Commission. It was given the major responsibilities of research, sea lamprey control and lake trout (Salvelinus namaycush) rehabilitation. This jointly supported agency does not have authority to manage and regulate fish populations; those functions are retained by each state and the province of Ontario; but it does serve to stimulate cooperative efforts throughout all aspects of fisheries research and management.

Sea lamprey control is an outstanding example of a diligent and sustained program effectively applying the tools produced by research. A lampricide (TFM) is quite specific to sea lamprey and has a reasonably short duration of toxicity. Therefore this serious parasite is controlled without adding to the contaminant burden of the Great Lakes.

The function to rehabilitate lake trout populations has had only mixed success. With the goal of a visible lake trout population sustained by natural reproduction, stocking begin in 1959 in Lake Superior where the sea lamprey first came under control. Stocking of lake trout followed sequentially in Lake Michigan, Lake Huron and finally Lake Ontario. The physical characteristics of Lake Erie (shallow, warmer and more turbid) presents a minimal potential for lake trout.

In Lake Superior, where substantial remnants of native lake trout remained, some progress towards self-sustaining status has been made. In Lakes Michigan, Huron or Ontario only a few instances of small numbers of naturally produced lake trout have been detected. There is no obvious reason why efforts to produce naturally spawning lake trout populations have been unsuccessful. A variety of contaminants in the lake trout and lake trout eggs may have been a factor in the early years when levels were several times higher.

As the lake trout and whitefish (Coregonus clupeaformis) populations collapsed in the 1950s commercial fisherman shifted to other species, principally chubs (Coregonus spp), yellow perch (Perca flavescens), lake herring (Coregonus artedii) and smelt (Osmerus mordax). As a result these species often were seriously over-harvested.

With the depletion of so many species of native fishes, and long before any control of sea lamprey any significant numbers of lake trout were restocked, the alewife (Alosa pseudoharengus) population began a very rapid expansion. The alewife had first appeared in the upper four lakes in the early 1930s. An anadromous species, the alewife had access to Lake Ontario but Niagara Falls had been a barrier until canals around the Falls were completed in the early 1800s. It remained unimportant until predators and competitors were reduced by the events previously described. The alewife populations began to make explosive-like increases in the 1950s. A short-lived species, it began an annual pattern of large die-offs following the stressful demands of spawning in late May and June. These die-offs assumed serious proportions in Lake Michigan and Lake Huron by 1957. Hundreds of miles of swimming beaches became littered with millions of dead and decaying alewife resulting in serious financial losses to tourist communities. The water intake screens and filters for the city of Chicago and steel mills were often plugged and damaged by dead alewives. By 1964 it was evident that the alewife made up over 95 percent of the biomass of Lake Michigan, and billions of dead alewives littered the swimming beaches each year. Present and common in Lake Superior and Lake Erie, it has never reached problem levels in those lakes.

Commercial fishing for alewives was encouraged but low values prevented this approach from achieving effective control. Perhaps the first clue to problems ahead appeared when failures in mink reproduction were traced to DDT in fish meal from alewife in their diet. Efforts to cleanup by trawling for dead and floating alewives, hand pickup and burying, and the use of bigger and better mechanical beach sweepers were expensive, frustrating and inadequate solutions. In the search for a better solution it became apparent that the alewife was both a problem and an opportunity. Without significant predacious fishes the alewife would continue to end up dead on our beaches. On the other hand, the alewife represented an enormous food source for large predacious fishes.

It was at this point that Michigan made some very major policy decisions that were later to be emulated by all other states and the province of Ontario.

The Michigan initiative had several facets. It was decided to begin an active management program consisting in large part of the stocking of fishes capable of utilizing alewife as their principal food. Michigan decided that restructured fish stocks would be managed principally as a recreation fishery (key value) and that commercial fishing would be relegated to a secondary role.

Neither the decision to manage for sport fishing and to relegate commercial fishing to a distinctly secondary role, nor the decision to stock Pacific salmon was easily arrived at. Commercial fishing interests were adamant in their opposition. Many of the scientific community opposed the use of any exotic species.

However, the decision to proceed with recreational goals and with Pacific salmon and steelhead trout prevailed. In December 1964 and January 1965 one million coho salmon eggs were obtained from the state of Oregon. These were reared to smolt size and 850,000 were released into Lake Michigan and Lake Ontario. The first of these were released into the Platte River, a tributary of Lake Michigan, on April 2, 1966.

The salmon program has been a spectacular success. In the late summer and fall of 1967 the first mature coho (*Oncorhynchus kisutch*) began to be caught. They were large, averaging 12-14 pounds, attractive and an instant favorite of the fishing public. In the summer of 1967, however, an enormous diet-off of alewife occurred in Lake Michigan. The chinook salmon (*Oncorhynchus tshawytscha*) was added to the coho, steelhead stocking increased, and by the early 1970s the sport fishery was booming. Control of alewives by the introduced salmon was achieved several years later.

Michigan's sport fishing program was gradually put into place through the 41 percent of the Great Lakes which Michigan owns and manages. To produce the optimum numbers of salmon and trout required a major expansion of the Michigan hatchery system, an expansion that has only recently been completed. In addition, Michigan has built many harbors of refuge, boat docking and boat launching facilities to serve the safety and service demands of an ever increasing number of boats. These choices by Michigan have been followed to a large extent by other states and Ontario.

PROBLEMS OF CONTAMINATION

By 1970 the early successes of the salmon program was generating a great wave of enthusiasm. The always important Michigan tourist business was expanding and shifting its capacity to serve the visiting fisherman and their families in a long list of port cities on Lake Michigan and Lake Huron. Many years of ever expanding sport fishing and increased benefits were clearly predictable. The shift in emphasis away from commercial harvest to recreational harvest was spreading rapidly to other states.

With other species, principally the walleye (*Stizostedion nitreum*), Ohio was emphasizing sport fishing on Lake Erie. New York and the urbanized portions of eastern Ontario were anticipating sea lamprey control so that the salmon programs could begin on Lake Ontario.

Into this era of euphoria the environmental awareness began to identify general and specific problems of pollution and contamination. Rachel Carlson's book, "Silent Spring" had specific examples for the northern midwest. Concepts of one world and the ecological reality that persistent toxic materials would move freely between soil and air applications to water began to be understood.

DDT was banned first in Wisconsin and Michigan and soon by both nations. We have learned to treat all other pesticides having the characteristics of persistence and broad scale toxicity very carefully.

As the salmon stocking increased many thousands of salmon began entering the spawning streams and Michigan began the harvest and sale of surplus salmon from state operated weirs. For several years these coho were marketed fresh, smoked and canned. Later DDT was discovered at levels above 22 ppm in coho flesh and the Michigan Conservation Department was ordered to cease its sale of salmon. Sales were halted but the Department continued to harvest salmon at stream weirs and to give them away to holders of a fishing license. After one year this was halted. For the next several years the surplus of coho and chinook salmon were buried at approved waste disposal sites. However, the sale of salmon eggs as preserved bait was permitted.

In 1970 mercury in dangerous concentrations was discovered in the fishes of the St. Clair River, Lake St. Clair, the Detroit River and western Lake Erie. Levels were particularly high in the species near the top of the food chain: walleyes, northern pike (Esox lucius) and muskellunge (Esox masquinongy). Michigan, Ontario and Ohio acted swiftly to close all commercial harvest of fish from the affected area and to sharply restrict their retention by sport anglers.

Most sources of mercury were quickly pinpointed as chemical plants using mercury electrodes in the production of chlorine from salt brines. It had been assumed that mercury lost daily in the water discharges would not be dangerous. We now know that some of the mercury was metabolized and moved up through the food chain to appear at dangerous levels in fish. The impact of mercury contaminated fish affected the livelihood of many people. Compliance with the regulations promulgated to prevent the consumption of mercury contaminated fish were frequently ignored. It was fortunate that once its use was abandoned an immediate drop in levels of mercury in fish occurred. In a matter of a very few years nearly all fish were judged to be safe. At the present time high levels are found only in the muskellunge.

In an interesting adjunct development the walleye populations of these waters doubled several times during the closure brought about by mercury contamination and the sport fishery for walleye grew enormously in response. With the exception of the Ontario waters of Lake Erie the commercial walleye harvest has now ceased.

As mercury and DDT levels began to trend downwards it became apparent that PCBs (polychlorinated biphenyls) were present at dangerous levels, particularly in those species that had high fat contents, were long lived, and at the upper trophic levels of the food chain. Salmon, steelhead (Salmo gairdneri) and especially brown trout (Salmo trutta) and lake trout had levels that prohibited their sale.

Sources of PCB's were identified as the hydraulic fluids in the large metal shaping plants, in electrical transformers, and a wide variety of commonly used materials. Laws were passed in Michigan and other states outlawing the use of PCBs for most purposes. Their use in certain closed systems is permitted. Similar laws restricting PCBs have been adopted by both the United States and Canada. A major cleanup of known point sources of PCBs was undertaken.

The downward trend of PCB concentrations in fish flesh began almost immediately following these actions. The Food and Drug Administration standard of 5 ppm was reduced in 1983 to 2 ppm (skin on fillet). At the present time all salmonids, with the exception of lake trout and brown trout over 50.5 centimeters and chinook over 64 centimeters, have concentrations well below the 2 ppm action level of the F.D.A.

Beginning with relatively small releases of lake trout into Lake Superior in the late 1950s, accelerating with the first introductions by Michigan of coho and chinook, the numbers of salmonids stocked has increased yearly. Each riparian state and the province of Ontario now have trout and salmon stocking programs. Many new hatcheries have been built, others expanded or modernized. The number of salmon and trout stocked annually in the five Great Lakes now exceeds 30 million.

There has been a shift away from coho and towards chinook, the popular choice of most sport fisherman. The numbers of steelhead planted has increased dramatically. The Skamania steelhead was first planted by Indians and more recently by Michigan. This large, hard-fighting trout has added one more dimension to the ever increasing popularity of Great Lakes fishing. Efforts continue to stock significant numbers of Atlantic salmon (Salmo salar).

A diversity of highly prized game fish have been selected to re-establish the predacious element of the fish species complex of the Great Lakes. The average size of salmon at maturity has been our best criteria of the adequacy of the stocking rates employed. In Lake Michigan the size of mature coho and chinook has declined substantially in recent years. In response, managers have reduced the numbers planted to about 14-15 million young fish. Lake Huron stocking does not appear to have reached the carrying capacity of the forage base and can be expanded further. Lake Superior stocking continues to be principally directed towards a goal of re-establishing a self-sustaining population of lake trout with salmon and steelhead of importance only in the areas where sport fishing predominates. Stocking rates remain stable.

Lake Erie is not well suited to the stocking of salmonids, however some expansion of stocking programs is underway and appears consistent with the forage available. Lake Ontario was the last of the Great Lakes to have an effective lamprey control program and to be stocked with salmon. The chinook, there, will occasionally still exceed 40 pounds and the population of forage species appears underutilized. New hatcheries and expanded programs will continue to increase the rates of stocking.

In summary, the stocking program of the Great Lakes has increased many fold from its early beginnings in the 1960s. Many millions have been spent to expand hatchery facilities and the capability to stock all the lakes at their optimum levels is probably now available. Fisheries managers can adjust the stocking rates annually to retain or improve the fit with the carrying capacity of each lake as displayed by increasing or decreasing size of the fish caught.

The restoration of balance between forage and predator fish populations has been achieved and will be maintained largely by the production of greatly expanded fish hatchery capability. That is not to say that there is no natural reproduction.

The program to re-establish the native lake trout has always had as its goal to recreate naturally reproducing populations in all areas where the lake trout was formerly present. Why there has not been a higher level of success is not known.

By the late 1960s and early 1970s naturally spawned young trout should have been present. When none could be found it was suspected that the high body burden of contaminants present in mature lake trout might be the problem. Mature lake trout in the early years of restocking were carrying body burdens of DDT in the range of 15 to 25 ppm and later carried similar levels of PCBs. Eggs collected from mature lake trout from the Great Lakes did experience higher than normal losses of eggs and sac fry. However, later as concentrations of contaminants were lowered abnormal losses have not been observed. Several other theories to account for the failure of the lake trout program to progress towards self-sustaining populations continue to be pursued. In spite of over 25 years of stocking effort these programs have yet to succeed in establishing any significant number of naturally spawned lake trout, let alone self-sustaining populations.

In the late 1960s coho eggs were first collected from Lake Michigan fish. Abnormal mortalities were observed and concentrations of DDT in the egg yolk seemed high enough to be the probable cause. As concentrations of DDT declined so did these problems.

Early in the plans to introduce Pacific salmon we conducted a review of the literature regarding their spawning requirements. From this it was apparent that while many of Michigan's streams were of suitable **quality** to support successful spawning, the **quantity** of stream areas would be too limited to provide the levels of recruitment necessary to achieve optimum population levels for the Great lakes.

The chinook are spawning in most available stream areas where water of good quality is present. It is difficult to estimate the total contribution of natural reproduction, but an estimate of 10-15 percent seems reasonable.

Natural Reproduction of steelhead has been increasing for several years and makes up a very significant but unmeasured percentage of the steelhead populations. The coho also reproduce successfully in a number of streams. However, numbers are lower.

The remaining Pacific salmon is the pink salmon (Oncorhynchus gorbuscha) and its presence in the Great Lakes is in the author's view an unsolved problem. A small number (estimated 20,000) of pink salmon were released from a Canadian hatchery on a tributary in Lake Superior where they were being reared for an experimental release in Hudson Bay. This unauthorized and unintended release occurred in the mid-1950s. Beginning on the subsequent odd-numbered years small numbers of pink salmon were discovered in Minnesota streams tributary to Lake Superior. By the early 1980s pink salmon were found in Lake Huron and in Lake Michigan. By 1985, they had reached Lake Ontario. The abundance of pink salmon is viewed with alarm by most fish managers. Anglers find them to be a nuisance. The province of Ontario allows a commercial harvest from Lake Huron of 300,000 pounds.

The food of the pink salmon is principally small smelt and alewives, placing them in competition with the most valuable coho and chinook and lake trout. Under present circumstances the unauthorized release of the pink salmon can only be viewed as an unfortunate mistake.

The restocking of the Great Lakes with an array of salmon, trout and other has been achieved and maintained with the annual stocking of many millions of smolts and fingerlings. Natural reproduction of the chinook has contributed substantially. Pink salmon are expanding totally by natural reproduction. The natural reproduction of steelhead, smaller numbers of lake trout and coho make an important contribution. This very large and expensive stocking effort added by these instances of natural reproduction has achieved a balance between forage fish and predacious sport fish. The sport fishery that has been developed has been a most spectacular success. The economic activity generated is estimated at $2 billion annually and still increasing. Several thousand anglers spend millions of days of fishing effort and enjoy leisure, fun and relaxation. In Michigan alone, Great Lakes sport anglers caught over 5 million salmon and trout in 1984. The fishing opportunity is and will remain of extremely high value because it has been created in waters accessible to millions of people. A high percentage of the populations of both countries are no more than one day's drive away from the fishing ports of the Great Lakes.

All of this has been achieved with the presence of previously described contamination problems. Actions have been taken that have reduced dramatically and quickly the levels of contamination. A review of what we have learned should provide the best available guidance as we make environmental and fisheries management choices for the future.

THE FUTURE

We have learned much about the Great Lakes ecosystems during the past 25 years. However, our knowledge is still imperfect, our facts incomplete and our response systems only partially effective. What follows will be a mixture of facts and opinion. An attempt will be made to make clear the distinction.

The Great Lakes ecosystem is large, valuable and vulnerable to mismanagement. Its water quality has frequently been lowered by man's past mistakes in waste handling and by the materials broadcast as pesticides. While some problems may originate from outside the Great Lakes watershed, most sources are within the watershed.

The processes of movement of toxic materials from one media to another; from soil to waters, from air to water, has been demonstrated in many instances. We understand that such translocations will occur. We understand that toxic materials may travel from great distances, i.e. toxaphene and DDT.

Within the lakes we have learned that while some portion of the problem materials will be taken out through sedimentation and absorption, other portions will be metabolized into the biological food chain. Contaminants will move upward from one trophic level to another with increased concentrations at each level. We know that fish will also acquire added amounts from the water through respiration and absorption and that the longer fish are exposed the higher the levels will be. Many contaminants, especially halogenated hydrocarbons, tend to concentrate in the fat.

In our efforts to control contaminants we know that by eliminating the use of i.e. DDT, for example, concentrations in fish flesh will drop rather quickly. In our Great Lakes experience declines of 75-80 percent occurred in four to five years. We know that elimination of point sources, i.e. mercury, can achieve equally dramatic reductions.

We know which contaminants and what types are likely to be translocated, bioaccumulated, and present unacceptable risks. We should be able to avoid many future mistakes rather than to have to correct them later.

It is clear that restriction or banning the use or the closing of point sources will reduce, but cannot eliminate contaminants from fish flesh. Ubiquitous materials will continue to arrive in the lakes from a variety of non-point and distant sources. As yet, we have no practical approach to achieve control of these latter sources.

We still have substantial disagreement as to what levels of what contaminants can be permitted in fish to be sold. Likewise, how stringent should our regulation be of recreational fish harvest and what exactly to say in our consumption advisories. We have had a good deal of confusion of what fish to sample, how many, and how to prepare for analysis (skin on or skin off fillets). However, we have recently achieved better coordination and better standardization of analytical procedures. It is also clear that the trend of opinion, as to what levels of the various contaminants are safe, is also toward still lower levels.

Much has been learned and achieved by the fisheries managers of Great Lakes in the 25 years. They have collectively been able to restore biological balance to a very large natural aquatic ecosystem. The original fish populations had been depressed, parasitized, species exterminated, and dominated by low value introductions. They took the view that they could manage and manipulate these unbalanced populations towards a productive fishery. To a very substantial degree they have achieved this goal. In so doing they made the choice to allocate these public fisheries to their highest value, recreational fishing. The previously dominant commercial allocation continues to be religated to a secondary role.

The newly created recreational fisheries are generating at least ten times the economic activity previously generated by commercial fishing. Further, the value of the recreational fishery will continue to increase in the years ahead.

The original concept of the ecologist, that the best management approach to the Great Lakes fish stocks would be to restore to the maximum extent possible, the native fish communities and to harvest from them the amounts of fish available from natural reproduction has been put aside. In its place the manager plants the species and the numbers of highest value fish (to the recreational fishery) that professional judgment and experience selects as most appropriate. This system has yielded the greatest economic and social values. Further, it provides the sustained capability to apply new facts and improved judgment to management strategies. By increasing and decreasing numbers of fish stocked, or by shifting emphasis between two or more species, managers can adjust to changes in the natural system, changes in public need or to take advantage of new information that promises to further increase the value of these public resources.

The fish manager has been joined by other professionals who add important dimensions to the value of recreational fishing. Ancillary to the fishing itself it is necessary to have adequate boating facilities and support services. In addition, since the fishing party is often a family group, it is important that the fishing port communities supply other leisure time opportunities for the non-fishing members of the party.

Anglers are going to harvest several million trout and salmon each year. They are willing to spend $50 to $75 dollars for every fish caught. This enthusiasm for fishing, the financial interest of the manufacturing and marketing companies, the income for services provided in each port community, coupled with the political strengths of these groups, provide massive assurance that the catch and consumption of trout and salmon from the Great Lakes will continue and probably increase.

The sport fishing public (and fish consumers) have been supplied with information through the news media, guidelines from state, and provincial agencies, and from warnings accompanying fishing licenses, about which fish are safe to eat without restraint, which ones should be eaten no more than once a week, and which should not be eaten at all. Pregnant women or those who expect to become pregnant are urged to follow even more restrictive guidelines for fish consumption.

The fishing public has been provided with directions on how to prepare fish for cooking in a manner that provides for the removal of as much fat as possible. The recommendations locate the area of fat deposits as belly, the back and beneath the skin along the laterial line. Preparing as fillets with the skin removed plus the trimming of all visible fat will lower the amount of contaminants substantially. In addition to these steps the form of cooking can also be important. Cooking methods such as broiling, which permit the fat to drain away during cooking, are also recommended.

In a review of experiences to date, there are some that should be emphasized. First, the level of all contaminants in Great Lakes fish have decreased substantially, in most cases by at least 75-80 percent. Nearly all salmon and steelhead examined since 1983 have been below the action levels established by F.D.A.. In a majority of these fish, concentrations are less than half the action level of F.D.A.. Further, these measurements are taken from skin on fillets. The vast majority of anglers, and the author has observed hundreds, prepare their trout and salmon as skin-off fillets.

Lake trout and brown trout continue to have concentrations of contaminants higher than other salmonids. Lake Ontario has higher levels of contaminants than the other four lakes.

The period of highest exposure by people to contaminants in Great Lakes fish is now many years behind us (early 1970s). Every reasonable expectation indicates that levels and risk will continue to decline. It is equally certain that we can never expect to reduce to zero the level of contaminants.

From past experience we should be able to define how to handle and use safely the materials that have the potential of creating problems in biological systems. We should have the energy and courage necessary to promulgate and enforce the laws and rules necessary in the handling of such materials. We must discontinue the manufacturing of certain materials such as DDT.

The state and federal agencies charged with the enforcement of these laws must be adequately funded and supported in a manner that assures that they will vigorously and effectively enforce the restrictions that are developed. Current levels of funding are inadequate and frequently declining.

One of the more difficult political tasks that lies ahead is to reduce air-borne contamination. The most frequently heard term, acid rain, is an important dimension but it is, in the author's opinion, too restrictive. The perception of the general public should be expanded to include all dimensions of air-borne contaminants. The arguments that we need to know more before attacking our air-borne contamination are in the author's view without merit. There is more than adequate knowledge and technology available. It is the political will that is lacking. The political difficulty is that one segment of people must bear the cost of clean-up and the benefits of cleaner air will accrue to people downwind in other states and provinces.

The ubiquitous "non-point source" will continue to leak a low level of contaminants into the Great Lakes. Reductions here must come from understanding and action from all of us. There must be fewer careless discharges and thoughtless applications. Accidents must be reduced.

In the application of pesticides we must further restrict the use of materials with the potential to be translocated. Alternatives to pesticides must be explored and made more financially attractive. When such materials are the only alternative we must encourage the use of minimum levels and minimum frequency of application.

The fisheries management agencies, often in connection with health and environmental agencies, can to some degree reduce the amount of exposure to contaminants by sports anglers through various adjustments and management regulations in stocking choices. Many of these adjustments probably would be unpopular and politically difficult. The Waukegun Illinois Harbor sediments contain a very large amount of PCB in high concentrations from past discharges from the Outboard Marine Corporation manufacturing plant. In the author's opinion this harbor and the adjacent portions of Lake Michigan should be closed to sport fishing. This would be a safety precaution but it would clearly put pressure to cleanup a serious source of PCBs that has not been solved because of legal deadlocks over how, by whom and how it should be paid for. However, it remains a serious problem to all of southern Lake Michigan. Similar closures should be considered whenever "hot spots" are identified.

The Federal Food and Drug Administration has the primary responsibility of setting action level standards for the various contaminants in food, including fish, and enforcements of those standards. Predictably the lake trout marketed from the upper three Great Lakes and Lake Ontario will contain concentrations of PCBs well over the action level of two parts per million. Most lake trout marketed legally are taken by various tribal groups under newly won treaty rights. The F.D.A. appears very reluctant to enforce standards of these fish when sold interstate. The same appears to be true of state agencies with similar authority over instate sales. The public has a right to be protected. Rules must be enforced.

Within the current array of trout and salmon being stocked by the various state and federal agencies, some will accumulate higher levels of contaminants than others. These, principally lake trout and brown trout, when caught and eaten, will mean the ingestion of more contaminants. This problem of higher levels of contamination in lake trout is recognized by anglers and visible in their slang references to lake trout as "sludge trout," "crisco trout," or "glow-in-the-dark trout." The author is well aware that there are legitimate management goals such as the elusive self-sustaining population of lake trout and the creation of local tourist attraction with brown trout. However, in the author's opinion the management agencies should carefully re-examine their commitments to these goals. Should these agencies deliberately rear and release species of fish that present higher levels of risk from contaminants to the angling public, especially when other species preferred by the public (steelhead and coho) that typically have safer levels of contaminants, are available? This question is worthy of the most serious consideration. It is a matter of serious ethical consideration.

Another alternative might be to enact regulations to enforce the release of all lake trout caught. This approach might serve the double purpose of reducing risk to the consumer and reducing pre-spawning mortality and thus enhancing the chances of establishing self-sustaining populations.

Anglers need more guidance. They should expect prudent advice from the health and environmental agencies regarding the consumption of fish. Warnings should neither overstate or understate the risks. Warnings, hopefully, can be made less restrictive if present declining trends continue. The present format of warnings should be expanded to provide advice as to how fish can be cleaned and cooked in ways that reduce contaminant levels in the portions to be eaten. The news media should convey information as to what geographic areas have higher concentrations of contaminants and anglers can then choose to fish in areas where contamination is lowest. Smaller fish as well as fish caught in the spring and early summer have lower levels of contaminants. Fish to be frozen can be selected accordingly. Anglers have an obligation to their families and themselves to follow the warnings as issued.

In the author's opinion the news media has consistently overstated the risks present. For example: In review of a report citing new data the media typically will select as examples the worst case; the species with the highest concentration from the area with the highest concentration to be followed with generalized discussion clearly implying that the examples cited are generally representative of all or most species, and present at those levels throughout all of the Great Lakes. Though the news media seldom makes a factual error. But frequently the treatment of subjects that raise levels of apprehension, such as fish contamination, are distorted leading to exaggerated conclusions by the reader.

CONCLUSIONS

Modern day fisheries management techniques, courageously employed by professional fisheries biologists, have succeeded in creating the world's most valuable freshwater sport fishery. This fishery has been recreated from the shambles of past management of native fish stocks, from a new mix of species, many of them carefully selected exotic species of salmon and trout, and in an environment abound by 100 years of abuse. This important and valuable fishery, with millions of people participating and worth over 2 billion dollars annually, has evolved in a period when a variety of contaminants have been a persistent risk and concern.

Many effective actions have been taken to reduce the levels of contamination in the flesh of Great Lakes fish. Most salmon caught by Great Lakes anglers today will be well below the established action levels of the F.D.A..

It is clear that some low level of contamination will be present for a long time to come. We have learned from 25 years of experience that all of us have roles to play in achieving further reductions in contaminants. We can never predict zero levels but we know that the highest levels of risk are years behind us. We can reasonably expect contaminant levels to continue to decline. With reasonable prudence the Great Lakes fishery can be enjoyed by millions.

CHAPTER 13

FISHES, FISHING AND POLLUTION IN LAKE VANERN (SWEDEN)

Bengt Lundholm

Eco Research and Resource Planning, Nockebyvagen 39, S 16140 Bromma, Sweden

ABSTRACT

Lake Vanern is the third largest lake in Europe. Once it was a good salmon lake, but industrial emissions and hydroelectrical exploitations have affected fish populations and fishing. In 1965 high mercury levels were found in pike. At the same time the pH in the lake was dropping and local eutrophication was recorded. Problems with fishing appeared. The commercial catch of salmon in 1970 was down to half a ton. Thanks to new laws and regulations those trends are now broken. Mercury levels are decreasing, the pH is slowly increasing and the water quality as measured by transparency is improving. Fish catches are also increasing, but the catches of salmonids are still far below the pre-industrial catches. The changes of the salmonid populations are of special interest as there are plans to increase the number of salmon to create a base for an intensive sport fishery, which can give new jobs which are desperately needed. Restoring the natural reproduction is difficult as the remaining populations are very small and possible reproduction sites are few. An increase of the salmonids in the lake must be based on stock from hatcheries. The present breeding stocks in the hatcheries is also very small and genetic investigations have shown that the genetic variability has been lost. We have now plans to protect and rebuild the local populations and if possible restore the genetic variability. It is, however, important that the smolts from the hatcheries are released in a way so they do not interfere with the natural populations. The goal is to re-establish Lake Vanern to the best salmon lake in Europe.

BACKGROUND INFORMATION

Lake Vanern is the third largest lake in Europe. The area is 5,550 km^2, which is less than a tenth of that of Michigan. It has a short geological history. During post-glacial times when the ice was retreating to the north there was a broad channel from the Atlantic Ocean covering Middle Sweden. When the land rose, this channel was cut off and a freshwater lake was created. In this lake two species of andromous salmonids were landlocked (Figure 1).

When the land rose the shores were covered with glacial clay, which is good soil for agriculture. Farming is now extensive on the flat lands in the South. On the broken ground in the North, forests dominate forming the base for an important industry. The watershed covers 46,800 km^2. The north-south length is 530 km and the east-west width is 180 km. In the northern part there are long rivers with many rapids and waterfalls. The longest is Klaralven - the Clear River. In the east, south and west are shorter, calmer rivers (Figure 2). There is a marked difference in the climate between the northern and southern parts. This was of importance when the salmonids adapted to the local conditions. This adaptation has however been destroyed by the erection of hydroelectrical power plants. The most important power plants are in Klaralven and in the river Gota Alv, which is the outflow from Lake Vanern.

Once 22 rivers produced smolt for the lake. Today only two rivers are left and the remaining populations are very small - so small, that they have been placed on the Swedish list for endangered species. The remnants of the once prospering salmonid populations are presented in Figure 3. Each river has thus two species. One, salmon (Salmo salar), the other brown trout (Salmo trutta). The species are similar. The local population and in the official statistics they are called "salmon". Sometimes the real salmon is called the "black salmon" and the trout is called the "gray salmon" or the "square tail". Since the Middle Ages salmon fishing has been of importance. In one early report from the governor of the northern province, it was stated that the province was very poor

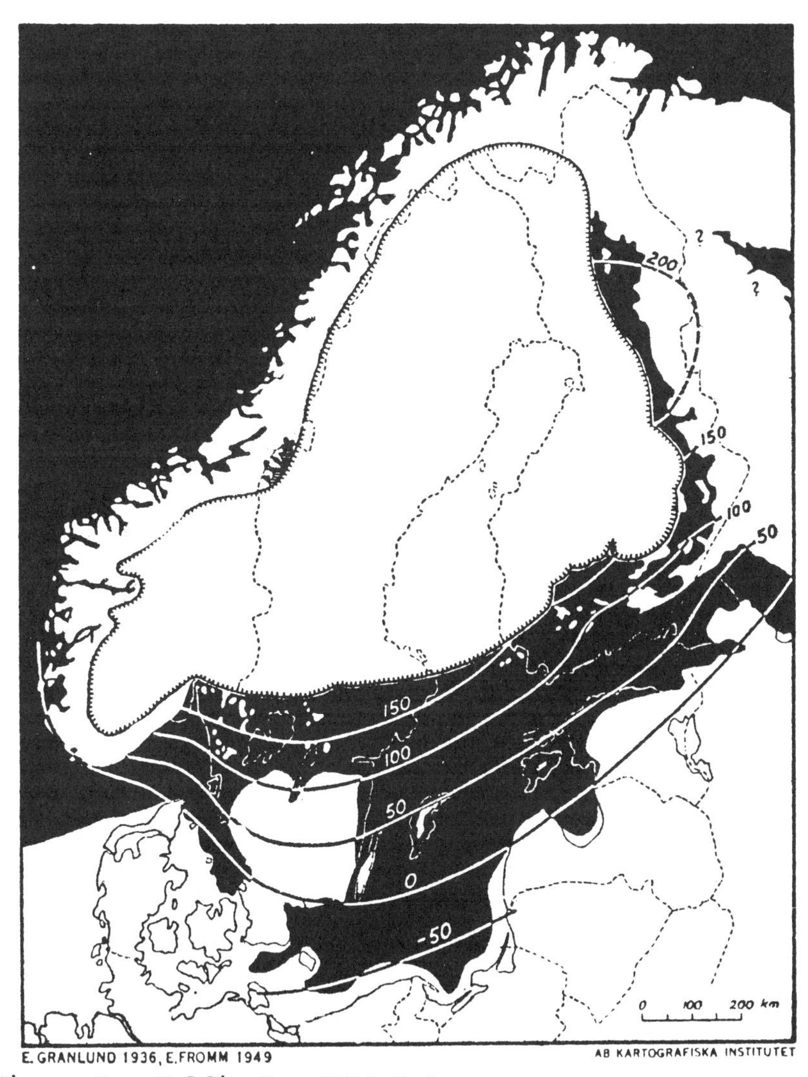

Figure 1. Yoldia Sea 7500 B.C.

and that the only asset worth mentioning was the salmon fishing in the river Klaralven. There were stationary fisheries which belonged to the Crown. During one summer in the middle of the last century 30,000 salmon were caught at one single location. The total amount caught in the lake and in the rivers has been estimated to be between 200 and 300 tons. Today the catches are around 50 tons, but in the middle of the seventies the commercial catch was down to 500 kg.

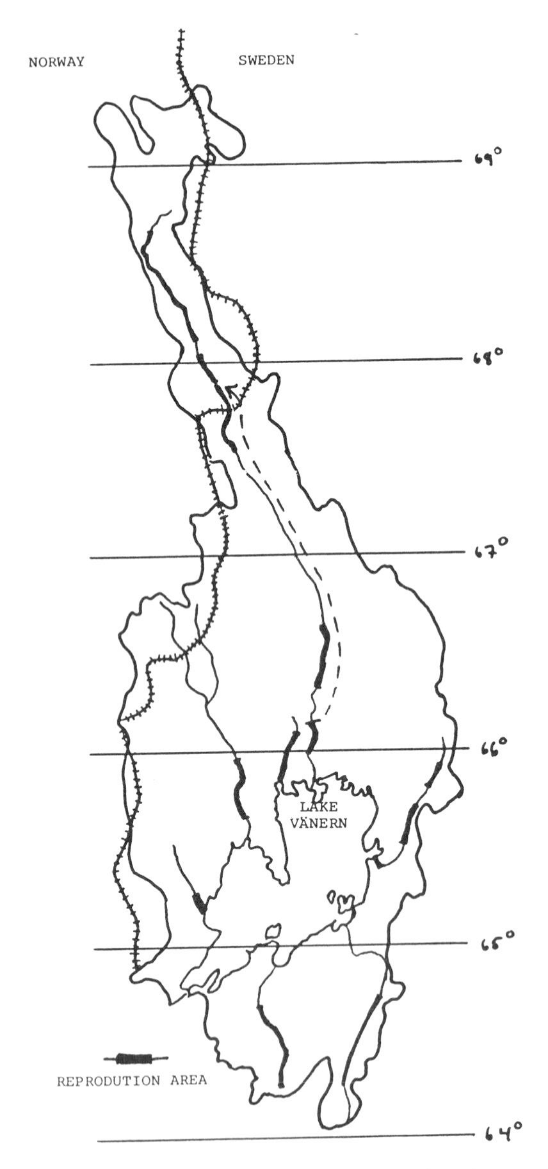

Figure 2. The watershed (arrow indicates transport of salmonids for breeding).

In this connection it is of interest to mention that sportfishing also occurred during the last century. It was introduced by the famous English big game hunter Lloyd. He used a rotating spoon and copper line. It was thus a kind of down rigger. Lloyd tells us that during his best fishing year he caught 195 trout in Vanern. Their weight was 850 kg with an average weight of 4.25 kg. From 1850 it is stated that in a day one could catch between 12 and 15 salmon weighing between 3 and 10.5 kg.

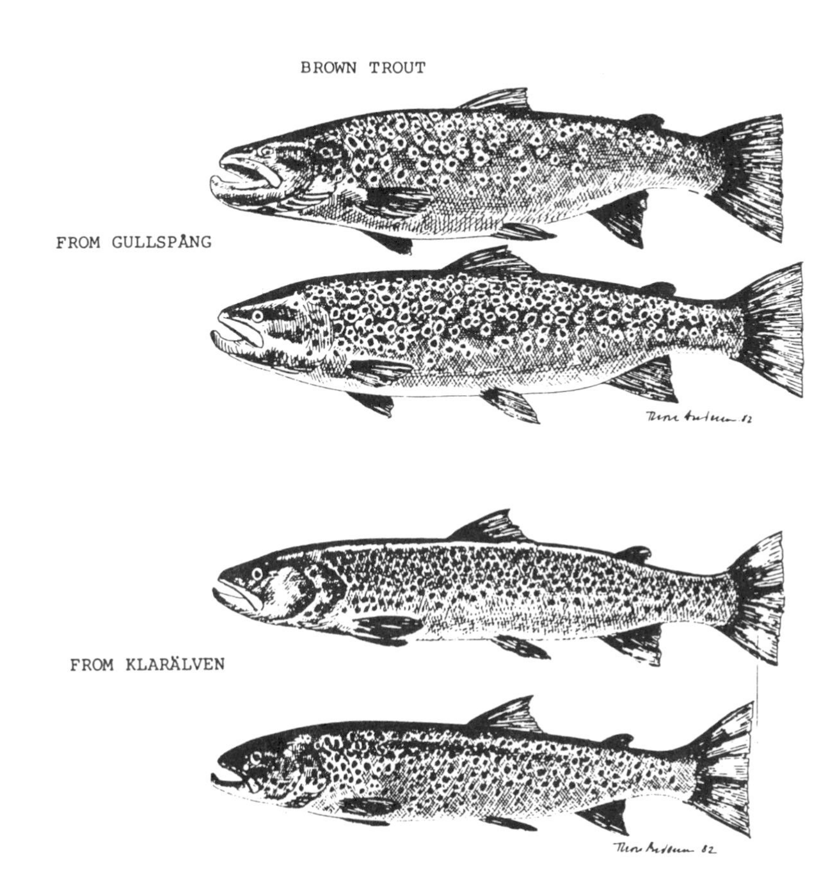

Figure 3. Brown trout.

THE MERCURY PROBLEM

In Sweden it was the mercury problem which made the public in general and the politicians in particular aware of the environmental problems. It actually started the environmental movement. Here the pike in Lake Vanern played an important role. In 1965 it was found that pike from the northern part of Lake Vanern had very high mercury content of in their muscles. At the same time it was shown that fishermen around Lake Vanern had high mercury levels in their blood. Later we found out that Sweden was on the brink of having a Minamata accident. Luckily, however, the Swedish government acted rapidly and forbade fishing in areas, where pike weighing 1 kg had mercury levels in muscles which exceeded 1 mg/kg. The thus blacklisted area in Lake Vanern was 280 km^2 (Figure 4). At the same time the emissions of mercury from the forest industry were strongly reduced. As a result of this "mercury scare" the prices on fish especially on pikes plunged as nobody wanted to buy them. A new market was however found in France and an export of pike started. In France pike, with it's white flesh was highly prized and there was no testing for mercury! Even if the emissions from the industry were reduced from 3,000kg before 1968 to 15 kg by 1980, the mercury had accumulated in the sediments (Figure 5).

After the reduction of the emissions the expected decrease of the mercury levels has been followed in detail in the most contaminated area since 1971. In 1981 it was below the critical 1 mg/kg value and the blacklisting was abolished. The reduction of methyl mercury is shown in Figure 6. The same has happened in the other earlier blacklisted areas and now no areas are blacklisted in Lake Vanern.

In Sweden we thought that the problems in connection with mercury were well taken care of. That has now proved to be a big mistake. We do not have mercury in Lake Vanern, but because of acidification, mercury has been released from the sediments in thousands of other smaller lakes in remote areas far away from any emission. Recently, the health authorities published a long list of lakes where fishing is forbidden. This will especially have an effect on sport fishing and tourism and will make fishing in Lake Vanern more attractive.

OTHER TOXIC SUBSTANCES

In the late sixties it was found that fish in the Baltic had rather high levels of DDT and PCB. This was especially the case with the fat species such as herring and salmon. In 1970 one sample of 10 salmon from the middle of the Baltic had DDT levels, in muscles, expressed in mg/kg in fresh weight varying

COMMERCIAL FISHERMEN IN LAKE VÄNERN

YEARS	NUMBER OF FISHERMEN	AVERAGE CATCHES IN TON
1914-1923	1700	0,5
1934-1939	650	1,3
1940-1945	1000	-
1965-1970	150	3,1
1984	214	5,4

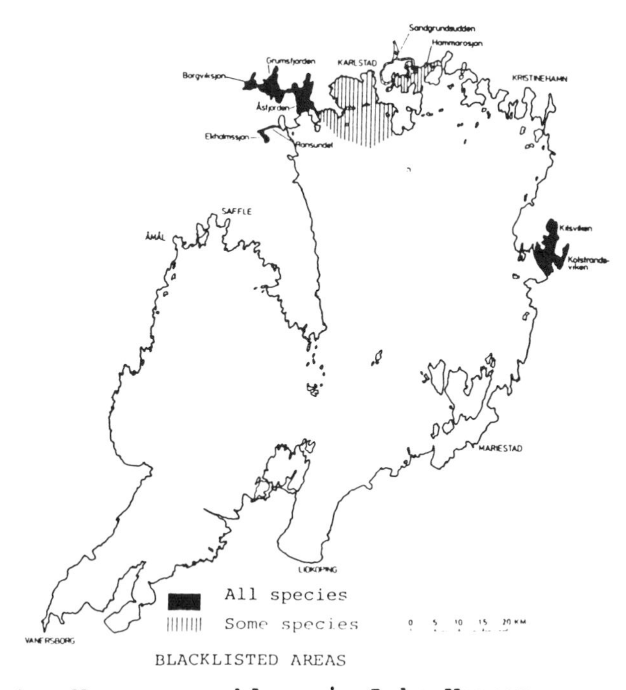

Figure 4. Mercury problems in Lake Vanern.

between 1.6 and 6.0 and the PCB levels varying between 1.1 and 2.1. On the Swedish west coast in the Atlantic the levels were lower. Since then the levels in the Baltic have gone down, especially DDT. In Lake Vanern the DDT and PCB levels in fish have always been lower than in the Baltic. Some recent analyses are presented in Table 1.

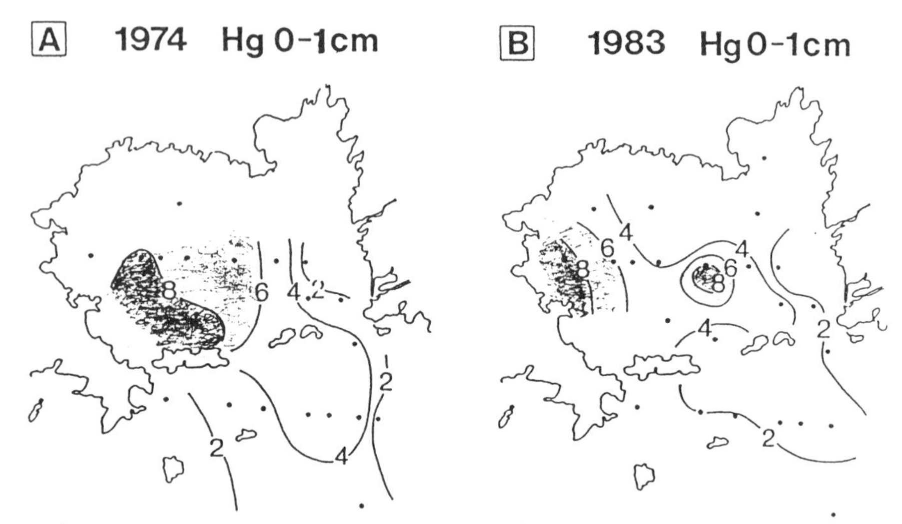

Figure 5. Total mercury content in surface sediments in northern Vanern (mg/kg dry substance) (after Lindestrom).

TABLE 1 LEVELS OF DDT-DDD-DDE AND PCBs IN SALMON (S) AND BROWN TROUT (T) FROM LAKE VANERN 1984

SPECIES	WEIGHT	FAT CONTENT	DDT-DDD-DDE	PCBs
	kg	%	mg/kg fresh weight	
T	0.7	2.6	0.04	0.10
T	0.8	2.1	0.04	0.09
T	1.0	3.8	0.06	0.15
S	3.5	8.9	0.19	0.53
S	3.9	7.3	0.16	0.47
S	4.5	4.6	0.13	0,35
S	5.1	7.3	0.26	0.72
T	6.3	4.5	0.15	0.40

According to the Swedish National Food Administration the maximum limits permitted in fishery products offered for sale in Sweden are 5.0 mg/kg for DDT-DDD-DDE and 2.0 for PCB's but for fish liver and salmon is 5.0 for PCB's. The levels found in Lake Vanern are thus low and present no health problems. This is important if we are planning for an increase of sport fishing in Lake Vanern.

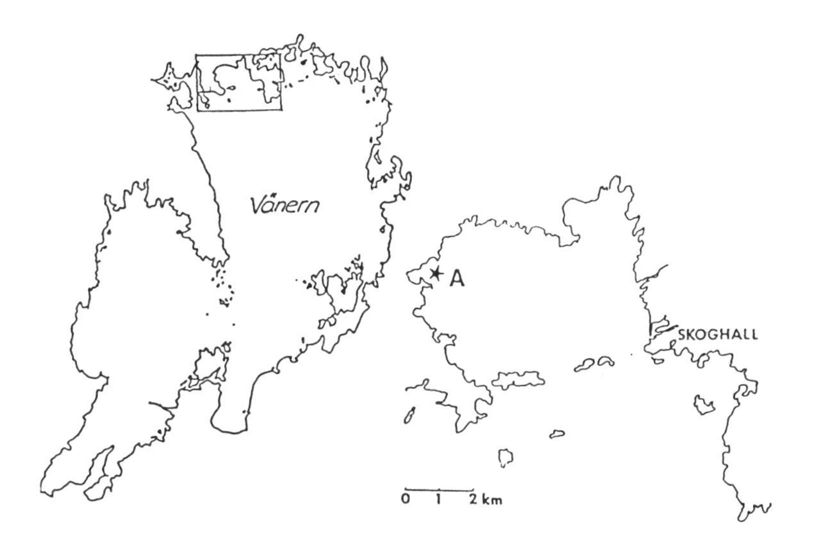

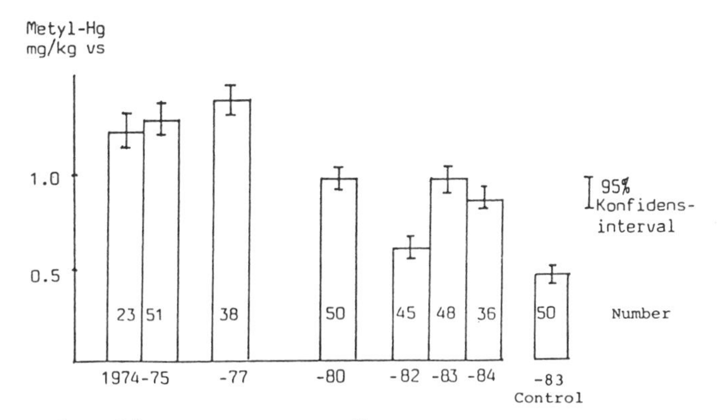

Figure 6. Disappearance of mercury.

ACIDIFICATION

The acid rain and other contributions to lake acidification are responsible for the release of mercury and metals, which have wiped out fishing in many lakes in Sweden and Norway. If we look at Lake Vanern we find a different situation. A pH graph of the water discharge from Lake Vanern is given in Figure 7. From 1965 to 1974 the pH dropped from 7.3 to 6.7

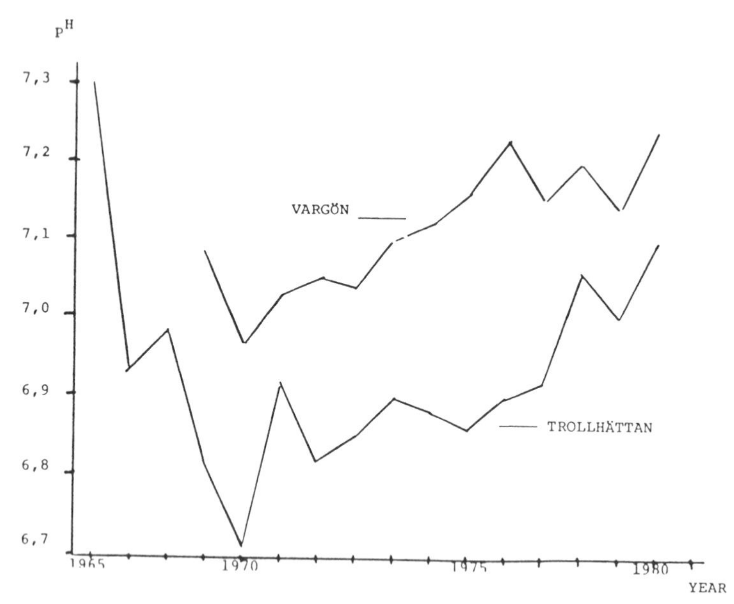

Figure 7. pH in the outflow from Lake Vanern.

but then the trend was broken and the pH has now increased. The main reason for this is that new environmental laws and regulations have reduced the acid emissions form the forest industries to the rivers. Figure 8 shows the increase of pH in the mouth of two northern rivers with big forest industries. It is also of importance that the large water volume in Lake Vanern is slowly responding to changes in acidification. For the present acidification is therefore not a problem in Lake Vanern.

FISH CATCHES AND POLLUTION

In the North Sea the commercial fish production is 60 kg/ha year, in the Baltic it is 35 and in Swedish inland waters, only 2.8 kg. The main reason for the low values for inland waters is neglect of commercial fresh water fishing in Sweden. We have very few fishermen using these waters. In Lake Vanern, however, commercial fishing has long traditions and is still handled by professionals. In spite of this the catches are low.

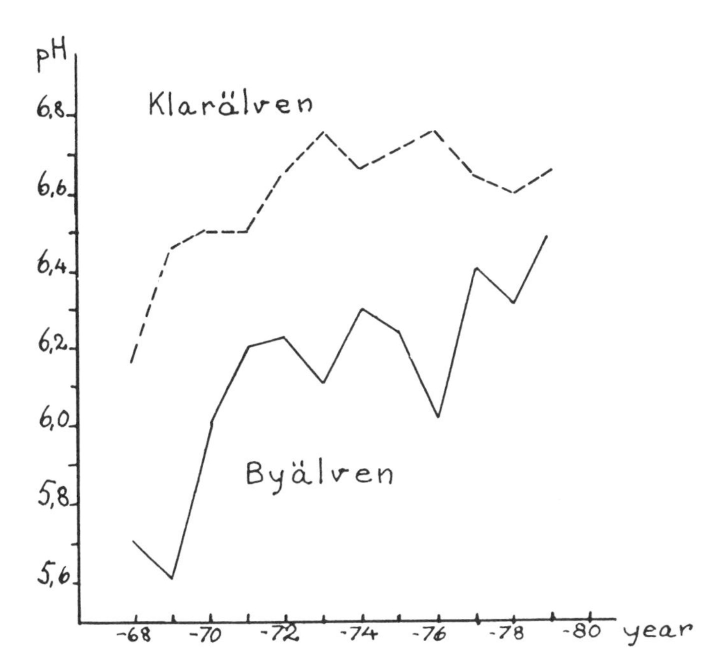

Figure 8. pH in northern rivers.

Fish production in a large lake is dependent on many factors and among other, is the relationship between surface and depth. A more shallow lake has a higher productivity. In 1975 Ahl developed a diagram, which shows that Lake Vanern compared with other large lakes has very low productivity when measured as commercial catches (Figure (9). This has been called "the Vanern paradox." It has been suggested that the explanation of this is that the forest industries discharge very dark pollutants, which had reduced sun light penetration. This in turn lowers photosynthesis and primary production.

Commercial fish catches are defined in different ways and depend on many factors, such as, market, fishing technique, catch regulations and changes in fish populations. Catch statistics from Lake Vanern show increased catches. Between 1962 and 1972 the average catch was 554 tons and between 1973 and 1984 it was 822 tons and last year it reached over 1000 tons. Measurement of transparency in the middle of the lake shows that the water is getting better. Sunlight is penetrating deeper (Figure 10). It is however doubtful

COMMERCIAL FISHPRODUCTION
HG / HA . YEAR

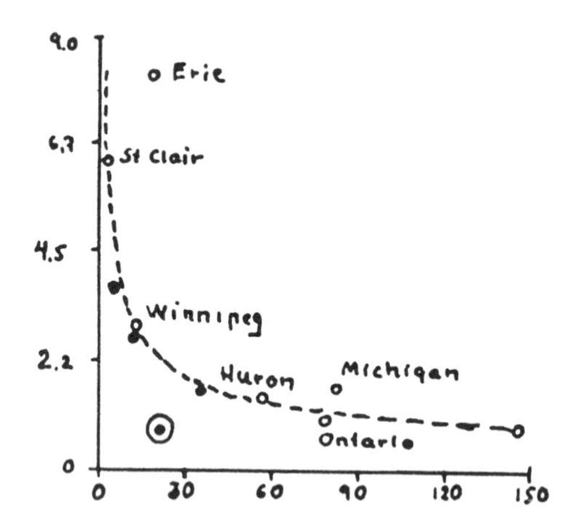

Figure 9. The Vanern paradox (from Rawson 1955, Ahl 1975).

if this alone can account for increased catches. Another explanation may be the increased eutrophication. Eutrophication occurs when nutrients released by man's activities have an effect on the different waters. From a human point of view it may be unfavorable. In such a case the nutrients are called pollutants. In other cases the addition of nutrients may be beneficial and give higher productivity and a better fish harvest: Everything depends on the state of the lake.

Vanern is an oligotrophic lake. A Vollenweider diagram shows that the human-induced annual nitrogen and phosphorus loadings have not reached dangerous levels in Lake Vanern (Figure 11). We have probably also had increased leaching of nutrients to the lake caused by the rising and regulating of the water level in order to get more hydroelectric power from the waterfalls at the outflow. This changed nutritional status has not only resulted in a higher productivity but it may also have given the lake a better buffering capacity against acidification. The increased productivity is even more pronounced when we consider the substantial catches made by a rapidly increasing number of sport fishermen.

TRANSPARANCY IN NORTHERN VÄNERN

	YEAR	TRANSPARANCY
STATION 1	1968	1,6 M
	1985	3,5 M
STATION 2	1968	1,7 M
	1985	3,5 M

TRANSPARANCY AT LURÖ IN THE MIDDLE OF LAKE VÄNERN

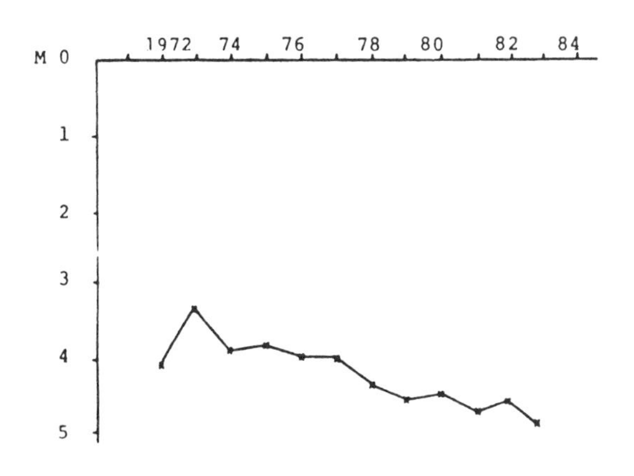

Figure 10. Increase in transparency.

THE PRESENT STATE OF THE SALMONIDS

As noted earlier, only two rivers produce smolts to Lake Vanern: Klaralven and Gullspangsalven. The present fish population in these rivers are remarkably different as shown in Table 2.

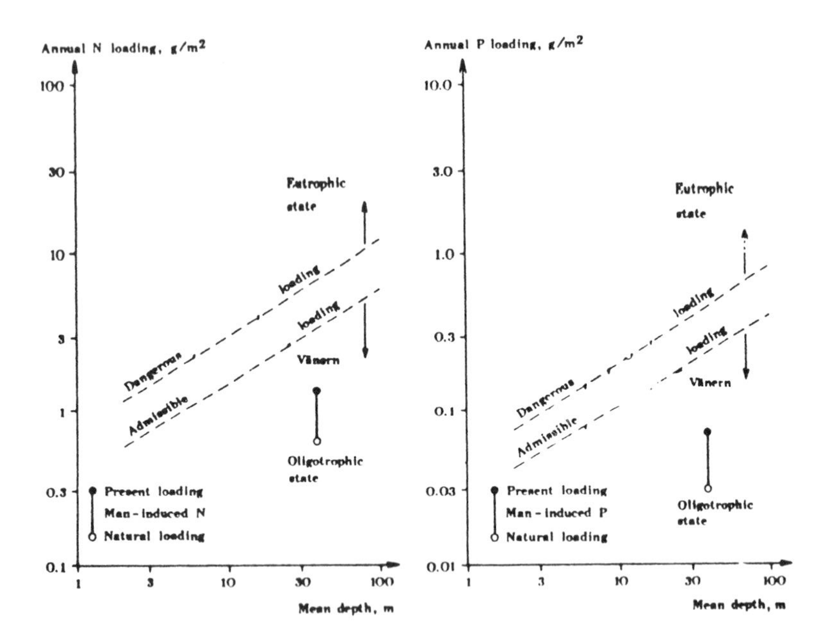

Figure 11. A Vollenweider diagram. The loading of N and P in relation to trophic level and mean depth in the Swedish Great Lake (from Vollenweider 1968 and Ahl 1975).

TABLE 2 CHARACTERISTICS OF THE REMAINING POPULATIONS OF BROWN TROUT AND LANDLOCKED ATLANTIC SALMON IN LAKE VANERN (after Ros 1981)

	Smolt		Adult						
	Age (year)	Length (mm)	Lake years	Spawn. weight (kg)	Migration starts	Migration distance (km)	Spawning time	Spawn (egg/l)	Max. weight (kg)
Atlantic salmon									
R. Klarälven	3	184	3	3.1	June – early Sept.	~ 350	Oct. – early Nov.	8,100	5
R. Gull-spångsälven	1–2	141–191	4–5	7.2	Oct. – mid. Nov.	74	Nov. – early Dec.	4,800	18
Brown trout									
R. Klarälven	4	235	2–3	2.2	Mid. May – early July	~ 400	Oct.	8,700	6
R. Gull-spångsälven	2	271	3–5	7.5	Oct.	~ 90	Mid. Oct. – early Nov.	5,400	22 ?

The most remarkable difference between the fishes in Klaralven and Gullspangsalven is in the growth rate. The fish from the last river grow much faster as is shown in Figure 12.

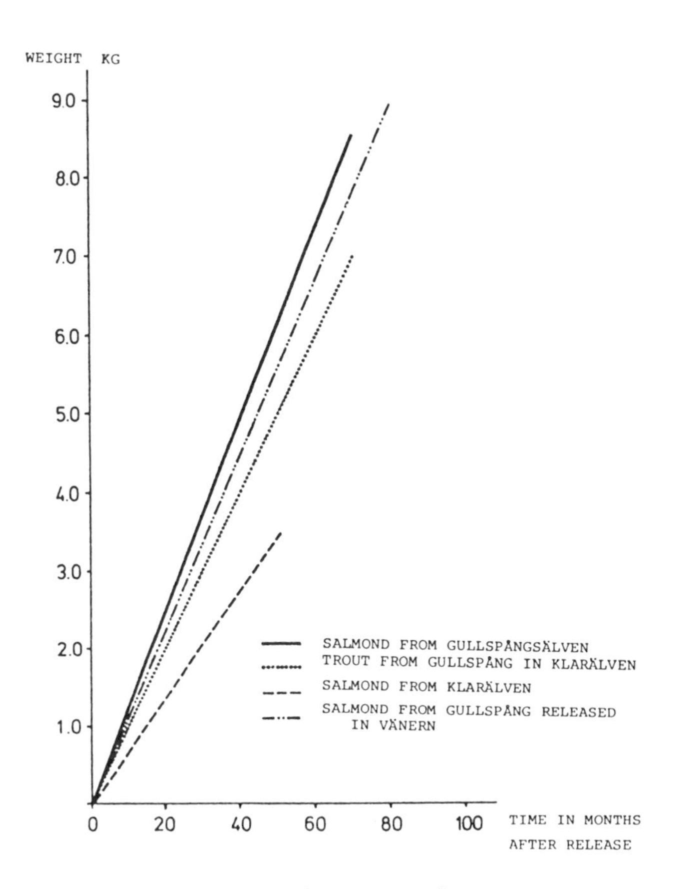

Figure 12. Growth rate in salmonids from Vanern.

In Klaralven now relatively small salmon remain with the spawning migration which starts in early summer (June) and with a long migration distance. The brown trout in the same river is larger with an even earlier migration start (May) and a longer migration distance. The spawning grounds of both populations are in the uppermost reaches of the river. The natural spawning pattern in the river has been completely disrupted by the hydroelectric exploitation. It may be assumed that once there were many different populations reproducing in this very long river. Short-migrating, large-size populations once lived near the mouth of the river, but these have now been exterminated.

The migration of the long-wandering populations was blocked by a series of hydroelectrical plants. The first is at Deje 15 miles from the mouth. There is a stationary salmon fishery owned by the power company. In order to close the salmon ladders and save water, the hydroelectrical interest were since 1933, allowed to transport by road a part of the migrating salmonids caught at Deje and release them above the power dams. The idea was to preserve natural reproduction. Nothing, however, was done to secure a safe way, for the smolt, down to the lake. The smolt have to pass all the turbines on the way down the river. Very few survive.

The power companies were also required to stock salmonid smolt in order to compensate for the damage to the salmon fishery. A certain part of the catch at Deje should have been used as parents for the hatchery smolt. The rest of the fish caught at Deje could be slaughtered. Also, the smallest fish were not to be slaughtered. Hatched salmon have been stocked in the Klaralven since 1960. The decisions and regulations of the water court were considered temporary while it awaited the results of studies. Now, we have results but the regulations are still temporary!.

The results of these regulations are now appearing. Let us first look at the number of fish returning to the river and being caught at Deje, where once 500 could be caught in a day or 30,000 in a year (Figure 13).

Marking experiments have shown a decreased fish return rate to the river. Between 1962 and 1970 1.1% of the released fish returned to the river and between 1974 and 1979 only 0.3% returned. In his extensive review paper, Stabell (1984) discussed homing in salmonids. He is very critical of the "imprinting" theory and favours the pheromone theory, which is that pheromones from related young or mating fishes guide the fishes back to the home river. He also stresses that there is a strong genetic element involved. Many experiments have shown that hatchery produced salmon have difficulties coming back to their home river. Stabell has pointed out that hatchery fishes have the same survival rate as those naturally produced and that the reduced homing may be explained by genetic break down.

In a large river there are several genetically different populations. In the hatcheries these are

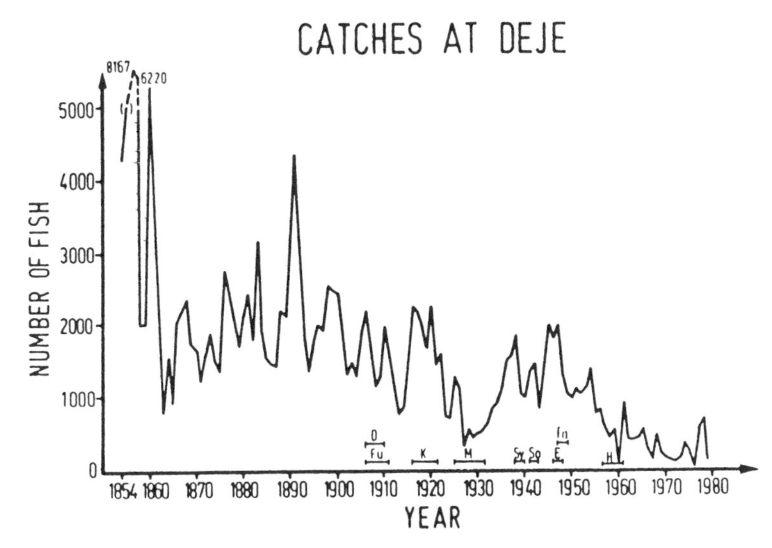

Figure 13. Catches at Deje (River Klaralven). The D, E, Fa, Fu, H, K, M, So and Sy indicate when differnt power plants were built (after Ros 1981).

mixed up. This has probably also been the case in Klaralven. More important, however, is that stocking in this river has been based on smolt from other rivers especially from Gullspangsalven. It has been claimed that the fishes from the two stock do not interbreed as one is short migrating and the other is long migrating. In hatchery producing smolt for stocking in Klaralven deliberate crossings have been made between salmon from the two rivers. This is probably the reason for the poor return rate to the river.

The return rate is not only of importance for the size of the fish catches in Klaralven, but it is also of importance for the general fishing strategy, but it is also of importance for the general fishing strategy. At the moment we are facing serious problems in relation to salmon fishing in the Baltic. There the fishing has been based on both stocking from hatcheries and natural reproduction. The proportion between hatchery and naturally produced material has been changed and the hatched fishes are more and more dominating. As a result fewer fishes are returning to the rivers. It is now difficult to get a sufficient number of parents for the hatcheries.

A consequence of the few returning fish is that far too few parents have been used in the hatcheries. When hatchery-rearing of smolts started, no one was aware of the importance to keep the genetic variability and genetic adaptation to the local conditions. Even if the opinion among the experts vary about the number of parents needed, for practical purpose the minimum number is 25 males and 25 females. Often these numbers have not been available or not been used.

Recent genetic investigations have proven that hatchery material of salmonids from Lake Vanern has lost genetic variability. This is considered to be a serious loss, since restocking of the lake has to depend, to a large extent, on hatchery material. However, it might be possible to restore the genetic variation by a careful use of the few remaining fishes naturally or semi-naturally reproduced. For several reasons - not only lack of material - Klaralven was been stocked with smolt from other rivers. The brown trout has especially been maltreated. As a result the original long migrating form has also disappeared.

In Gullspangsalven the situation is different. The natural reproduction system in the river has received very little interference. Initially, a stretch of 74 km was available for salmon migration. In 1906 a power station was built 8 km from the mouth. A salmon ladder was built, but that was closed in 1924. Although part of the remaining reproduction sites were illegally destroyed as a result of bottom cleaning done by the power company, the river contains populations of both species of brown trout. In 1940, S.Runnstrom described the remarkable fish population in Gullspangsalven. There are still naturally reproducing fishes of both species left in the river. They have been well protected.

Material from Gullspangsalven has been taken to different hatcheries where breeding stocks have been kept. From these hatcheries material has been widely distributed. The genetic investigations have shown that genetic variability has decreased and that there are genetic differences between the different breeds. Hatchery material has to a great extent been used for stocking in Lake Vanern but the main part of hatchery stock has been put out in Klaralven. We hope that the hatchery material has not interfered with the remaining naturally reproducing populations.

As indicated in Table 3, the stocking of smolts has given very good returns for the fishing in Lake Vanern.

TABLE 3 STOCKING OF SMOLTS AND COMMERCIAL CATCHES IN LAKE VANERN

PERIOD	COMMERCIAL CATCH tons/year	STOCKING OF SMOLTS number
1964-1966	3.9	22 400
1967-1970	3.3	10 400
1971-1974	1.4	37 500
1975-1978	14.2	78 400
1979-1981	37.0	61 900
1982-1984	41.6	

As mentioned before there are no reliable statistics for the catches made by sport fishermen. During the last years there has been an increase in sport fishing and quite a few salmon are caught in the lake. According to the estimate for 1984 the 1620 registered non-professional fishermen caught 14.2 tons and the 214 professional fishermen caught 51.8 tons of salmon and brown trout.

Compared with the fish caught in Lake Michigan the fish caught in Lake Vanern are small. I happen to have information about catches from one of the few boats using a deep rigger in Lake Vanern (Table 4).

TABLE 4 CATCHES USING A DEEP RIGGER IN LAKE VANERN

YEAR	FISH number	AVERAGE WEIGHT kg
1981	97	1.85
1982	23	2.43
1983	41	1.83

This low weight, compared especially with the records from the last century, does not depend on lack of food. One of the main prey species is the cisco (Coregonus albula (L)) which is fished for its roe. The catches of cisco have been increasing, from 85 tons in 1976 to 426 tons in 1984. The growing salmonid populations have not reduced the cisco catches. The reason for the rather low weights is probably the very high number of gill nets, which catch the young salmonid. In the lake there are now more than 30,000 gill nets!

THE FUTURE

There are now plans to increase salmonid populations in Lake Vanern. The target is to reach catches of 200 to 300 tons by releasing a large number of smolts, in order to increase both the commercial and sport fishing of the lake. An important argument is that sport fishing is expected to create new jobs, which are badly needed in the areas around Lake Vanern. In order to achieve this goal, many things must be coordinated. Proper genetic material has to be found, or even created. New hatcheries have to be built. Fishing regulations have to be changed. New arrangements have to be made to receive new tourists interested in fishing. Special marketing has to be done. Most important, money for all this has to be found. The problem is that it will take several years before the lake is filled with salmonids, but we will try hard to create the best fishing lake in Europe.

REFERENCES

Ahl, T. 1975. Effects of man-induced and natural loading of phosphorus and nitrogen on the large Swedish lakes. Internationale Vereinigung fuer Theoretische und Angewandte Limnologie 19.

Andersson, O., Linder, C.E. and Vaz, R. 1984. Levels of organochlorine pesticides, PCBs and certain organohalogen compounds in fishery products in Sweden, 1976-1982. Var Foda. 36:suppl.1

Hakanson, L. 1981. A manual of lake morphometry. Springer Verlag, New York.

Lindestrom, L. 1984. Mercury in sediments from Lake Vanern. (Swedish, Mim).

Lundholm, B. 1975. Interactions between oceans and terrestrial ecosystems. In: S. Fred Singer (ed.), The changing global environment.

Ros, T. 1981. Salmonids in the Lake Vanern area. In: Fish gene pools (ed.) Ryman), Ecol. Bull. (Stockholm) 34 pp. 21-31.

Stabell, O.B. 1984. Homing and olfaction in salmonids. A critical review with special reference to the Atlantic salmon. Biol. Rev. 59 pp. 333-388.

Statens Naturvardsverk 1978. Lake Vanern - a natural resource. (Swedish).

Statens Naturvardsverk 1981. Monitor 1981. Acidification in soil and water. (Swedish).

Vollenweider, R.A. 1969. Scientific fundamentals of eutrophication of lakes and flowing waters, with particular reference to nitrogen and phosphorus as factors of eutrophication. OECD<DAS/CSI, Paris.

CHAPTER 14

THE ADMINISTRATION OF THE FISH CONTAMINANT PROGRAM IN A DECENTRALIZED STATE AGENCY

Larry A. Nielsen

Department of Fisheries and Wildlife Sciences, Virginia Tech, Blacksburg, Virginia 24061 USA

ABSTRACT

Isolated problems with chemical contamination of Wisconsin fish require the Department of Natural Resources to issue fish consumption advisories for recreational anglers and other fish consumers. Because the department is decentralized, many groups share responsibility for the advisories. The absence of clear roles for these groups in the past has resulted in poor internal communication, non-rational data interpretation and contradictory public statements. A year-long evaluation of the program has produced a new operating procedure designed to assure both agency-wide review of data and decisions, and public announcements that are timely and consistent. The new procedure is a 4-stage process that features redefinition of the sampling year, annual analysis of data (except in emergencies), and extensive liason between field and staff and between fishery and water resource managers.

INTRODUCTION

The presence of toxic chemicals in fish flesh has become increasingly common over the past two decades. In Wisconsin, the well-studied contamination of Lake

Michigan fishes by organochlorines stimulated questions about contamination in other waters. Carp and catfish from the upper Mississippi River also have been found to contain high levels of organochlorines. A more recent problem is the discovery of mercury contamination in predatory fishes living in small, isolated inland lakes. This problem is particularly troublesome when the impossibility of adequately sampling thousands of small Wisconsin lakes is weighed against the necessity of protecting the health and recreation of more than 2 million anglers.

Fish contaminant problems generate intense interest because fishing is often an important local industry. Fishery management agencies must balance a primary objective of protecting human health with other objectives that are also critical. Closing fisheries or advising people to reduce fish consumption affects the income of commercial fishermen, charter boat operators, and resort owners. Agency statements that are complex or frightening may cause people to ignore the warnings and may even cause people to stop believing what the agency reports. Furthermore, the fundamental question of what constitutes a health risk still remains unanswered. In the midst of these scientific and social issues, the fishery management agency must use the best available information and procedures for making decisions (Milbrath 1983).

In Wisconsin, the Department of Natural Resources (DNR) is responsible for regulating fishing and for advising the public about fish contamination. The DNR has operated since the mid-1970s as a decentralized agency, with significant responsibility for decision-making and program operation vested in field offices. Decentralization has specific implications for the success of a fish contaminant program and all programs that combine local and state-wide interests and that require the interaction of different specialists. The objective of this paper is to describe the DNR's administrative structure as it pertains to the fish contaminant program, the specific problems that a decentralized agency has in issuing a fish consumption advisory and the evolving procedures that govern the program.

DECENTRALIZATION AND THE FISH CONTAMINANT PROGRAM

Agency Structure

The Wisconsin DNR is directed by a seven-member Natural Resources Board appointed by the governor. The chief executive officer is the secretary, also appointed by the governor. Six district directors have responsibility for major regions of the state and report directly to the secretary (Figure 1). Within five of 6 districts, three area directors have responsibility for smaller regions and report directly to their district director. Technical personnel (e.g. fishery managers, water resources managers) within an area report directly to the area director. This constitutes the line of authority in the department.

Staff specialists provide support for the secretary and district directors. The secretarial staff perform state-level work, including administration of the fish contaminant program. The extensive secretarial staff is organized into bureaus (e.g. Bureau of Fish Management, Bureau of Water Resources Management) which are the most visible units within the DNR. Technical staff also operate at district level, providing regional expertise to the district directors. Because of the small staff size, many district staff specialists oversee several technical areas.

Implications of Decentralization

Decentralization offers one major benefit for managing complex problems such as fish contamination. The benefit lies in the opportunity to assemble local personnel into an inter-disciplinary team. In this case, the relevant technical personnel include fishery managers and water resources managers, who can direct fish sampling based on local knowledge of likely contaminant sources, water flow patterns, fish movements, and fish harvesting practices. The team can apply the same expertise in later stages of data interpretation and decision making. This advantage also extends to higher levels, where staff to the district director and secretary can interact directly so that coordinated and fully informed recommendations are carried forward to the directorate. The line-staff organizational structure embodied in decentralization is designed specifically for situations like this in which information flow outside of the simple line authority maximizes agency performance (Bozeman et al. 1978).

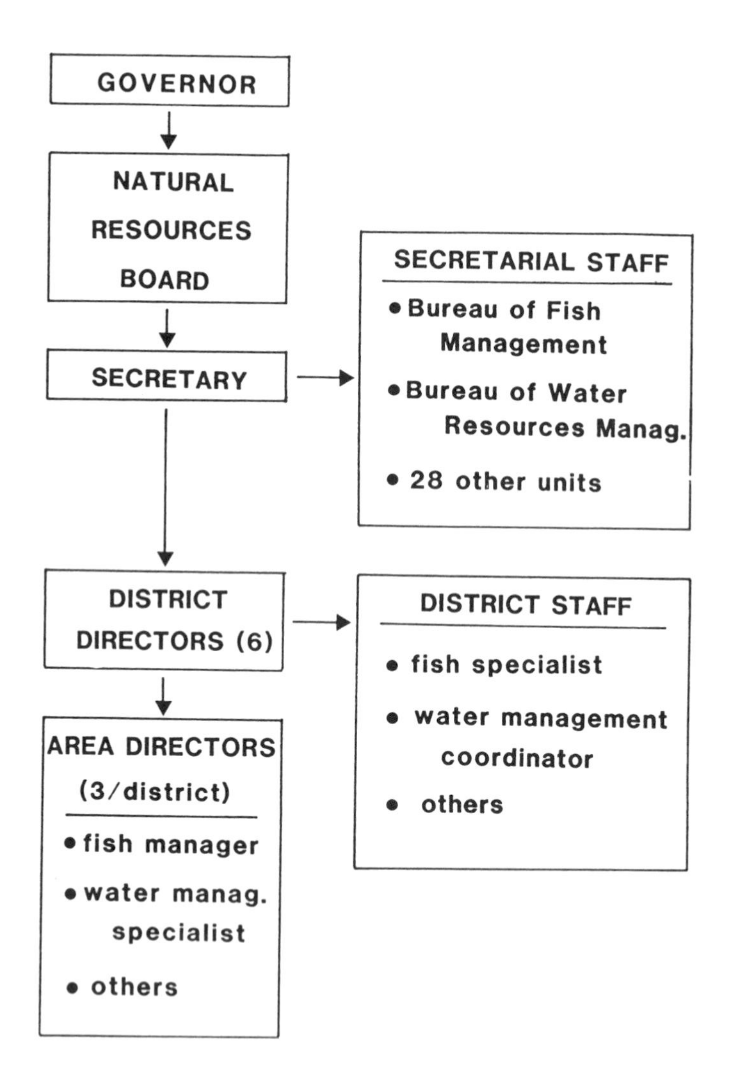

Figure 1. Organization of the Wisconsin Department of Natural Resources (line authority is on the left and staff functions are on the right).

The theoretical advantage of decentralization, however, is offset by difficult practical obstacles. The principal obstacle is conflict generated by the multiple viewpoints and centers of responsibility within the agency (Drucker 1954). The fish contaminant program had grown within the DNR from several independent origins to the point that those interests overlapped extensively. For example, fishery managers and water resource managers have fundamentally different approaches to fish contamination. Fishery managers are more interested in defining contamination of fishes important to the angling public and the commercial industry. Water resources managers are more interested in sampling fish where toxic contamination is probable, whether or not fishery is important. Fishery managers are likely to underestimate the severity of contamination relative to water resources managers. Similar differences in orientation are probable between line officers, who are responsible for specified geographical areas, and state-level staff, who are responsible for specific technical domains.

The second major obstacle to an effective decentralized program involves information flow after decisions have been made. Although administrators may encourage staff to exchange technical information informally, they generally adhere to strict hierarchical flow of information relating to policies (Peters and Waterman 1982). Policy decisions within agencies, therefore, are generally communicated slowly and often incompletely from top administrators to field personnel and from one decentralized location to another.

In contrast, communication through the media is nearly instantaneous. A press release to state capitol reporters will be transmitted by wire service to local newspapers and radio stations within the day. As a consequence, local agency staff may be contacted for comment on an agency policy well before they have seen the official policy. This has been a continuing problem for the Wisconsin fish contaminant program because the issues generate intense media interest at both state and local levels. The general public, furthermore, does not necessarily understand or respect decentralization. They expect local fish and water resource managers to be fully informed about local and state-wide decisions as well as decision affecting distant localities where they have interests.

The problem of effective communication is amplified in Wisconsin by an active press and sophisticated public interest groups. These organizations utilize freedom-of-information statutes to access information at all levels. Decentralization, which spreads information widely in an agency, creates more potential outlets for the demanded materials. The information releases, beginning with the raw laboratory data, may develop into a continuous stream that the agency cannot anticipate or duplicate through internal communication.

THE WISCONSIN FISH CONTAMINANT PROGRAM

The Wisconsin fish contaminant program illustrates the realities of administering a complex program amid intense public interest and within a decentralized agency. In the remainder of the paper, I describe the program as it existed through 1984 and as it has been revised for future years. The observations are based on my year-long special assignment to assess interactions among fisheries and environmental units sharing responsibility for specific DNR programs, including the fish contaminant program.

The paper focuses on procedures for program operation rather than on technical decisions; such items as what constitutes a suitable sample, what qualifies as an emergency, or what toxicant levels define a health risk are not addressed. The procedures describe the pathways for actions, rather than the substance of the actions. These are both important policy domains, but procedural matters are primary because they provide the framework for rational technical decisions.

The 1984 Program

During and before 1984, the fish contaminant program operated informally and without comprehensive procedural guidelines. Personnel in the Bureau of Water Resources Management (BWRM) initiated the program each year by developing a fish collection schedule based on the previous year's data and on emerging water quality concerns. After minimal review by fisheries staff in the Bureau of Fish Management (BFM), the schedule was sent to field fish managers, requesting them to collect and process fish according to specific instructions. In the absence of both field review and strong apparent commitment by the BFM director and their area directors, fish managers frequently neglected the program. They collected fish only as time permitted and often ignored instructions regarding sizes and species of fish needed and appropriate handling techniques. Fish collection was scheduled by calendar year (January to December), but most fish were collected in the fall when Lake Michigan salmonids returned to spawning rivers.

Fish were transported to the central office in Madison, where BWRM personnel processed the fish as time permitted for chemical analysis. Laboratory analysis was frequently delayed because of higher priority assigned to other programs such as water well contamination or toxic spill investigations. Most samples had been analyzed and data were available for interpretation by late March or early April.

Data interpretation was informal in most regards. No pre-determined guidelines existed for data assessment. For example, the difference in risk between two sets of samples, both with similar mean toxicant concentrations but with very different ranges, had not been established. Therefore, a team consisting of the responsible physician from the Department of Health and Social Services, BFM staff, and BWRM staff met at intervals to make decisions based on new data, new scientific opinion, and changes in policy. Because of other priorities, such meetings were generally scheduled as external demands for interpretation arose (e.g. renewed interest by the press, a printer deadline).

The primary mechanisms for releasing fish contaminant information was the "fish consumption advisory," contained in the annual fishing regulation pamphlet. The pamphlet serves well because it is distributed with fishing license applications through established pathways and because anglers actually read the regulations pamphlet. The pamphlet is less than optimal because only brief information can be involved; the typical advisory consisted of three 9 x 15 cm pages.

An additional limitation of regulation pamphlets has been the extended printing schedule. The printer's deadline is mid-October for materials appearing in the pamphlet issued in January, when the new year's licenses go on sale. Because the sampling year operated from January to December and because data became available in the spring, one cycle of the fish contaminant program lasted two years. Fish collected in 1982, for example, would be analyzed in late 1982 or early 1983. Data would be interpreted during 1983 for inclusion in the fishing regulation pamphlet issued in January 1984.

In 1984, therefore, the DNR began issuing a separate advisory, appearing as a single-sheet bulletin just before opening of spring fishing seasons. The purpose of the spring advisory is to incorporate data accumulated during fall and winter, an interval when many data become available. The spring advisory is problematic, however, because interpretation is non-systematic and driven by deadlines and because no distribution network exists. The DNR has used press releases to publicize the new information and to tell anglers how to acquire a copy of the advisory.

The Modified Procedure

In August, 1984, the U.S. Food and Drug Administration lowered the action level of PCBs from 5 mg/L to 2 mg/L. In response to this change and continued high PCB concentrations in carp from Green Bay, the DNR issued an emergency order closing the Lake Michigan carp fishery. Commercial fishers, anglers, and field fishery managers all objected to the emergency action for a variety of reasons, most related to administration of the program. The DNR, therefore, undertook a study of the fish contaminant program. The goal was to modify and clarify existing procedures to assure both accurate and timely action by the discovery of toxic contamination in fishes. After extensive review by field and central office personnel, a modified procedure has been enacted as departmental operating policy. The fundamental changes include the designation of specific program leaders in the Bureaus of Fish Management and Water Resources Management and the declaration of a detailed 4-stage process (Table 1).

TABLE 1 MODIFIED PROCEDURE FOR THE FISH CONTAMINANT PROGRAM (see text for definition of abbreviations).

STAGE	PRINCIPAL RESPONSIBILITY	REVIEW	TIME PERIOD
1.Collection schedul.	BFM staff	BWRM staff	Mar.-June
2.Data generation	field staff	BFM staff	June-July
3.Data interpretation	BFM staff	BWRM staff field staff other agencies	July-Aug.
4.Policy development	BFM dir. DNR sec.	BWRM dir. field staff	Aug-Oct.

Collection Scheduling

The first stage defines the specific schedule for the coming sampling year. The schedule is initiated jointly by BFM and BWRM program leaders, based on state-wide concerns and results from the previous year's sampling. This information is sent to field personnel, who add regional concerns and provide advice on sampling strategies for addressing state-wide concerns. BFM and BWRM program leaders then draft a complete sampling schedule. The schedule includes size, sex and species of fish to be collected; collection locations and times; explicit directions for storing fish and transferring them to the Madison laboratory; and chemical tests that will be performed. The schedule is reviewed by relevant personnel before being finalized and distributed by the BFM director. Approximately four months, March to June, are allocated for this stage.

Scheduling in this manner has several distinct advantages. **First**, the process utilizes knowledge of local managers to determine the most appropriate sampling program. **Second**, it enhances cooperation in fish collection by managers because their participation develops local "ownership" of the program. **Third**, it provides a specific step for interaction between BWRM staff, who process samples, and fishery managers, who collect fish, to ensure that methods are understood and accepted. **Fourth**, complete scheduling allows for an annual collection program that is compatible with time and financial constraints of all participants.

Data Generation

Data generation includes all logistic steps from fish collection to compilation of the annual data set. This stage is coordinated by district fisheries staff and the BWRM program leader. District fisheries staff oversee fish collecting, labelling and freezing of fish and transportation of the fish to the Madison laboratory. The BWRM program leader receives fish, processes them for laboratory analysis, transports samples to the lab, receives and initially reviews data and supplies data to computer operators for entry into the comprehensive data system.

This stage includes two significant alterations from the previous procedure. First is a redefinition of the sampling year to begin on July 1 and end on June 30.

Redefinition in this way moves the mass of fish collecting (i.e. in the fall) to the first half of the sampling year, providing a more even flow of processed samples to the laboratory. Because few fish are collected in the spring, the laboratory can catch up with the backlog of fall samples. By the end of June, virtually all data, even for fish collected during June, should be available for interpretation.

The second difference relates to the orientation towards an annual data set. Data interpretation (stage 3, below)) is suspended until the entire year's data are available in order to avoid inappropriate short-term decisions. The objective is to reserve judgments until temporal, spatial and species-specific trends in the data can be observed from an entire data set.

The procedure does include a provision for immediate action in extreme cases. The BFM and BWRM program leaders review incoming data regularly to identify unusual situations; they refer such cases to their bureau directors. Candidates for immediately action include samples from a particular water body, all of which exceed specified contaminant concentrations, or some of which exceed specified concentrations by a substantial amount. Separate protocols have been prepared to define what constitutes such exceptional circumstances.

Data Interpretation

After the entire year's data are compiled the procedure moves into the third stage, data interpretation. Interpretation is defined as characterizing the data in regard to the sampling and numerical characteristics. For example, interpretations must be made concerning what water body area a set of samples represents, whether or not two data sets can be combined, and whether or not separate species can be treated as a group. Interpretation is guided by pre-established standards created jointly by BFM and BWRM personnel and by the designated Department of Health and Social Services physician. The BFM program leader prepares an initial interpretation which is reviewed by a team of relevant secretarial staff. The revised interpretation is then transmitted to field personnel for review. The final interpretation is issued by the BFM director. The procedure allocates 2.5 months for this stage, ending in mid-August.

A distinct procedure is also defined for interpreting the data from a water quality perspective. The BWRM program leader and director perform identical roles to those of the BFM personnel described above. By separating the data interpretation stage, the fishery and water quality programs retain clear responsibility and leadership for their areas of primary interest.

Development of an explicit data interpretation stage solves a major difficulty in the fish contaminant program. As described earlier, time constraints and pressures for public disclosure often limited the DNR's capacity for thorough data analysis. By designating a specified period for specific responsibilities, the procedure institutionalizes rational consideration for the complete annual data set, including extensive internal review. The procedure also provides a definition of "information" for use in addressing freedom-of-information concerns: information exists only after data have been interpreted. Although this definition of "information" probably will be challenged legally, the establishment of a procedure definition provides a rational basis for the agency's position.

Policy Development

The final stage involves making decisions based on the previous data interpretation. Responsibility shifts at this stage to the agency administrators. The BFM director drafts recommendations for policy change and other actions and routes those recommendations to DNR district directors and staff administrators and to relevant personnel in other agencies (e.g. State Department of Agriculture, U.S. Food and Drug Administration). The BFM director revises his recommendations and forwards them to higher administrative levels or takes direct action, as defined by departmental authority.

The procedure allocates two months for this stage as it applies to fish consumption advisories. The availability of the revised advisory coincides with the printer's October deadline for the next fishing regulation pamphlet. (The spring advisory is eliminated under this procedure.) The information is also delivered to the public information officer, who designs and implements public information activities in conjunction with secretarial and district staff. This staff is designed to assure that DNR staff know about decisions and underlying rationales at least as soon as the public knows about them.

DISCUSSION

The modified procedure for the fish contaminant program has several advantages over the previous procedures. **First**, the explicit procedures increases the likelihood that program personnel will **act** rationally rather than **react** defensively. The existence of step-by-step operating procedures for the program institutionalizes rationality and shifts the burden of justification to those individuals and circumstances that demand suspension of the procedure.

Second, the new procedure shortens the time from first fish collection to issuance of a consumption advisory from 2 years to 1.5 years by redefining the sampling year. The redefinition especially increases the speed with which spring-collected fish are included in the decision-making process. The 25% reduction in program length by this change indicates the value of explicit and objective program evaluation. Many improvements in operating effectiveness and efficiency result from such common sense adjustments that they may not be apparent to those closest to a program (Bozeman 1979).

Third, the procedure mandates more complete internal communication before the agency takes any actions. Effective communication is essential for an interdisciplinary program like fish contaminant, especially so in a decentralized agency. The specific review procedures will surely be violated many times, but explicit mandating review institutionalizes cooperation as the spirit of the program.

The modified procedure, however, does not cure all ills associated with such a program. Forecast among the residual programs is the ever-present demand for instantaneous information. Everyone, within and outside the DNR, remains dissatisfied with the speed of transforming fish samples into agency policy. The deliberate pace of the agency seems unavoidable to most BFM and BWRM staff, but the public and field managers are much less patient. As a consequence, the procedure is already being modified to re-instate the spring fish consumption advisory. The spring advisory will be based on data generated and analyzed since the previous October, and the fishing regulation pamphlet will alert anglers to look for the separate advisory. Re-instating the spring advisory will compromise the rational basis for the procedure. It illustrates the power of urgency in public agency decisions.

The procedure also ignores the difficulty of actually interpreting the data. Technical judgments of what does and does not constitute a health risk are highly subjective and are likely to remain so. The high cost of a truly reliable sampling program also will continue to complicate decisions. In the absence of a national priority for resolving these problems, differences in interpretation will continue to divide both the public and the professionals who share responsibility for such programs.

Undoubtedly, the greatest value of the modified procedure, and of the study which lead to its development, is the increased attention devoted to cross-program cooperation. A year of constant attention to inter-disciplinary environmental programs heightened the DNR's sensitivity to the benefits of interaction. Liason responsibilities have been assigned to specific people in the relevant bureaus and information cross-program communication has increased substantially. The specific procedures are unimportant. They evolve continually as experience grows. Establishing a commitment by the people to interact continuously and openly is the key to success.

ACKNOWLEDGEMENTS

This study was supported jointly by the W.K. Kellogg Foundation, the Wisconsin Department of Natural Resources and Virginia Tech. It was possible only because of the candor and professionalism of the people of the Wisconsin Department of Natural Resources, especially James T. Addis and Bruce J. Baker.

REFERENCES

Bozeman, B. 1979. Public management and policy analysis. New York: St. Martin's.

Bozeman, B., Roering, K. and Slusher, E.A. 1978. Social structures and the flow of scientific information in public agencies: an ideal design. Research Policy 7:384-405.

Drucker, P.F. 1954. The practice of management. New York: Harper and Brothers.

Milbrath, L.W. 1983. Public decision-making with regard to managing major natural resources. Renewable Resources Journal 1(4):18-23.

Peters, T.J. and Waterman, R.H., Jr. 1982. In search of excellence. New York: Warner.

CHAPTER 15

FRESHWATER FISHERIES MANAGEMENT AND POLLUTION IN BRITAIN: AN OVERVIEW

K. O'Hara

Department of Zoology, University of Liverpool, Brownlow Street, P.O. Box 147 Liverpool, L69 3BX

ABSTRACT

The major centres of urban development in Britain are situated on rivers and the principal large lakes are in rural areas. Consequently the most serious effects of toxic pollution in freshwaters have been exhibited in rivers, with amelioration measures being directed at these problems. Freshwater fish in Britain are divided into game (salmonid) and coarse fishes, but total only fifty-four species. Coarse fishermen normally return all captured fish alive to the water body. In most intensively angled trout lake fisheries, fish are stocked as hatchery-reared, table-size fish and the majority of trout fisheries are of good water quality. Other than Atlantic salmon, sea trout and eel fisheries, only minor commercial freshwater fisheries for food exist, with very few in lakes. Human health problems associated with consumption of freshwater fish are therefore not perceived as a major concern. The different types of stillwater environments found in Britain are compared and their utilization by anglers described. Fish kill incidents have been calculated from contrasting areas and the occurrence of toxic induced problems, including those associated with acidification, are assessed. These are contrasted with other types of pollution. Climactic conditions strongly influence recruitment in coarse fisheries, causing considerable natural fluctuations and with relatively few recreationally important fish species diseases can also be important. Information sources available to the public and anglers are assessed and two case studies, acidification and lead poisoning in swans are used to illustrate fisheries management related problems.

INTRODUCTION

Britain being a heavily populated, industrialized but at the same time a geographically small country, has suffered and continues to suffer from pollution of some of its inland waters. As example is the River Mersey which has been described as containing a powerful amalgam of discharges and remains still one of the worst polluted rivers in the country (Holland and Harding 1984). Although water qualities are generally good, particularly in the large lakes, there has been, and remains, a keen awareness of the influences of pollution on fisheries and the existing freshwater angling groups reflect the impact of pollution on fisheries and the existing freshwater angling groups reflect the impact of pollution on fishing methods, species captures and attitudes to conservation. The present paper attempts to outline the nature of fisheries in Britain, some legislative aspects of fisheries management and pollution control. The preservation of fishes cannot be separated from the general theme of environmental protection and conservation issues. Public and angler awareness of environmental problems in aquatic ecosystems continue to grow and information sources which contribute to this knowledge are widespread and readily accessible.

LAKES IN BRITAIN - SIZES AND DISTRIBUTION

Considered on a world scale Britain does not possess any really large lakes, the largest in terms of surface area being Loch Lomond, Scotland (71 km^2). The largest lake in England is Lake Windermere (14.8 km^2). Scotland possesses easily the greatest number of lakes in Britain having 69 percent of the totals of standing water and the lakes are important in a European context (Maitland et al. 1981b).

The large British lakes are nearly all glacial in origin and consequently are often of great depth. Their locations in the North and West of Britain are in areas of low human population density and they consequently are of good water quality. As described later, standing water bodies are of great importance for recreational fisheries in Britain but in order to illustrate this importance small water bodies must also be discussed. Some of these smaller water bodies, particularly where they occur in groups such as the Cheshire/Shropshire Meres and the Norfolk Broads are of great biological interest.

Water demands for industry and drinking water have resulted in the construction of many inland reservoirs and the modification of some natural lakes. These artificial reservoirs can match and exceed in size natural lakes. In Wales Lake Vyrnwy has a surface area of 4.5 km^2 and in England Rutland Water has an area of 12.6 km^2. The man-made water bodies have quickly established themselves as major centres of water based recreation. Many of them are important trout fisheries maintained by artificial stocking (Moore 1982).

FISH FAUNA AND FISHERIES

The fish fauna of Britain is depauperate in terms of the number of species comprising 54 species of which 42 are indigenous Maitland (1985). This relative paucity of species is a direct result of extinctions during the last glaciation and the limited re-colonization by non-diadromous species that occurred before Britain became separated from mainland Europe. Some species, for example the common carp (Cyprinus carpio) and the rainbow trout (Salmo gairdneri) were subsequently introduced and have established in the case of carp or are stocked in the case of rainbow trout, and form the basis of important fisheries. These introductions would not be possible now under current legislation. Nevertheless angling represents the largest participant sport in Britain with a total of 3,380,000 anglers in England and Wales and 354,000 in Scotland; 8 per cent of the total population (NOP 1980). These anglers are grouped into three categories, sea anglers, game (or salmonid) anglers and coarse fishermen who fish for non-game species. Coarse fish species would include the pike (Esox lucius) and the percids perch (Perca fluviatilis) and zander (Stizostedion lucioperca), but the most popular and most frequently captured species is a cyprinid, the roach (Rutilus rutilus) (NOP 1971). An important difference between coarse and game anglers is that the former rarely retain their catch for consumption preferring instead to return fish alive to the water body. The relevance of this habit to the maintenance of coarse fish stocks has been questioned in some quarters but in heavily fished waters and/or marginal quality waters it is no doubt essential to avoid over exploitation. Multiple captures of the same fish have been frequently recorded. Cooper and Wheatley (1981) estimated that all fish in the River Trent were caught at least once annually and O'Hara (unpublished) has shown that in some fishing competitions catch rates of approximately 30 kg/ha can be achieved in a five hour competition. A general overview of fisheries in England and Wales is provided by Chandler (1981) and Edwards (1985).

In spite of man-made changes in the distribution of fishes through deliberate introductions, considerable differences remain in the number of species recorded in different areas of Britain (Maitland 1985). As a very broad generalization, fish communities change from cyprinid dominated in the southeast to salmon, charr and whitefish in the north and west, with perch, pike and trout having a fairly widespread distribution. Climatic and topographical conditions may not be conducive to the establishment of cyprinid populations in some areas and eurythermal species such as the common carp have to be sustained by stocking in many waters, particularly Northern Britain. In 1970 when the last detailed national survey of angling took place, rivers exceeded all other sites in importance for coarse fishermen, 46 percent as compared with 40 percent using still waters and 14 percent canals. Important regional differences were found according to water qualities and sites available. For example in North West England only 17 percent of coarse anglers fished on rivers because they are mainly polluted in the important Mersey catchment, whereas 40 percent of the anglers in this area fished on canals (NOP 1971). In 1980, 45 percent of game anglers fished on still waters (NOP 1980). Such variations influence management policies including pollution control because of the general assumption that non-salmonids are more resistant to pollution than the more sensitive salmonids (Solbe et al. 1985).

Because the larger lakes tend to be in areas of low population density and distant from urban conurbations they are not intensively angled compared for example with coarse fisheries located in or near to cities, and the large lowland English trout reservoirs. The 1980 NOP survey recorded that most coarse fish anglers usually only travelled a distance of 15 miles from their homes and most game anglers 30 miles. Therefore the rural-based natural fisheries are likely to have less angler pressure than will be experienced by urban situations and the professionally managed and stocked reservoirs. Maitland et al. (1981) for example, only record in the order of 1,000 anglers on the large Scottish lakes. This is in spite of Loch Lomond having a large number of species of potential recreational interest. Llyn Tegid in North Wales accommodates of the order of 4,000 anglers per year. This can be compared with Rutland Water's 47,000 angler visits in 1979 (Moore 1982) and coarse fisheries where hundreds of anglers may fish in one competition in a single day. Interestingly, Goldspink (1983) considering the Cheshire Meres, noted that quality of sport and rank order of preference was inversely proportional to the size of the lake, that is the smaller lakes produced the best coarse fish sport. Swales and Fish (1986) have noted that catch per unit area of trout is high in smaller reservoirs, this is probably due to lake morphometry and the fact that angling is mostly carried out from the shore. Only limited management of most of the natural lakes takes place with some stocking of trout e.g. Loch Leven to bolster natural stocks. The stocking of the larger coarse fish water bodies is normally not cost effective and most rely on natural sources of recruitment (Goldspink 1983).

However, in addition to any fisheries potential, these large natural water bodies are of important conservation interest because of their diverse fish faunas and the presence of rare and potentially endangered species such as the whitefish (Maitland 1985) and are considered to be of national importance.

Commercial fisheries were once an important feature of inland waters in Britain, for example Windermere (LeCren et al. 1972). Few commercial fisheries now exist other than for the migratory salmonids, the Atlantic salmon (*Salmo salar*) and the sea trout (*Salmo trutta*). These fisheries are mostly coastal or estuarine based intercepting returning adults, but there are commercial net fisheries on some Scottish Lochs, e.g. Loch Lomond (Maitland 1981). Eels (*Anguilla anguilla*) are commercially fished on some lakes but mainly on rivers as they descend as silver eels. Consideration of the potential accumulation of pollutants is not a major human health problem for coarse fish which are the main inhabitants of lowland rivers, where the principal toxic pollutant loadings occur (Solbe et al. 1984) and most trout originate from and are stocked into waters of good quality.

LEGISLATIVE ASPECTS OF FISHERIES AND POLLUTION

The control of aspects of water management in England and Wales is by Regional Water Authorities. Under the terms of the 1975 Salmon and Freshwater Fisheries Act they are required to:

> ..."maintain, improve and develop the salmon, trout, freshwater fisheries and eel fisheries."...

Pollution legislation is both of domestic and European Economic Community (EEC) origin. A full and competent review with regard to fishes is provided by Graham and Pearce (1981). The major development since that time has been the implementation in 1985 of the Control of Pollution Act (1974) Part 2, which importantly requires Water Authorities to provide a public register of water qualities and discharges.

POLLUTION SOURCES, WATER QUALITIES AND FISHERIES

As indicated previously, most major sources of pollutants tend to be centered on rivers and mostly on the lower and estuarine reaches. A National Classification Scheme of all rivers is operated by the Regional Water Authorities in England and Wales which recognizes different categories according to their water quality status and ability to support fish communities of a salmonid or cyprinid nature (NWC 1980). The close proximity of the major urban conurbations to the sea has resulted in a considerable amount of disposal of pollutants to the marine environment. Because of the widespread consumption of marine- as opposed to freshwater-fish, environmental health surveys take major cognizance of this source (MAFF 1986).

Recently, many Water Authorities are reporting that, among their major water quality problems, are discharges of an agricultural origin, particularly silage liquor and slurry. These organic spills often have a very high concentration of ammonia. Most incidents appear to be restricted to rivers. A continuing problem is the use of fertilizers which is causing enrichment problems in both rivers and standing waters.

The larger British lakes although influenced by enrichment have not suffered major deleterious changes.

Industrial discharges are not a problem. Enrichment of some of the important lake complexes however has caused major declines in biological quality. These have been admirably documented on the Norfolk Broads by Moss (1983). His paper convincingly demonstrates the complex biological interactions that can result from eutrophication. Similar concern has been expressed over some of the Shropshire-Cheshire meres (Reynolds 1974). However from both sites the data on fishes is much more limited than for other groups. One difficulty of assessing any changes in fish populations and communities in Britain, particularly for non-salmonids is the lack of available catch statistics and the absence of yield data because there are no commercial fisheries. Only relatively recently has the collection of data from non-game anglers been developed as a means of following population and community trends.

Among the principal toxic pollutants that cause concern are pesticides and metals (Mason 1981). As previously indicated, this is indeed the case for the marine environment around Britain (MAFF 1986). There is no major routine monitoring program for these pollutants in freshwater fish flesh in England and Wales although a pilot study was conducted between 1980 and 1983 on organochlorine residues from sites in central England. A Pesticide Precaution Scheme is operated by the Ministry of Agriculture, Fisheries and Food which evaluates the likely effect of new and existing pesticides on fish.

Radioactive discharges to aquatic systems are confined to coastal sites except for one site, Trawsfynydd Lake in North Wales, a man-made lake. Annual reports are produced on the levels of radioactivity found in rainbow trout, brown trout and perch which are caught by anglers.

This is not to suggest that problems may not exist in freshwaters but mostly they are localized and are likely to be noted where a fish kill has been observed. In once incident concerning stocked trout a warning was issued to avoid consumption of fish (Hamilton 1985).

Metal pollution is a feature of some water courses, particularly in the older industrialized areas and where old mine workings exist. Again it is not associated with our major lake environments but it is a problem in some coastal marine areas.

A comparison of the distribution of fish kills in the North West Water Authority Area is shown in Table 1 as an illustration of the difference between rural and

TABLE 1 NUMBER OF FISH KILL INCIDENTS RECORDED IN NORTH WEST WATER AUTHORITY 1976-1984, INCLUDES MORTALITIES BY DISEASES

Northern Division		Southern Division	
Rivers	Lakes/Ponds	Rivers	Lakes/Ponds
110	5	164	102

No recorded mortalities in Lake District lakes except Grasmere and Windermere, both with Perch disease.

industrialized areas. The Northern region contains the Lake District of which perhaps the most famous lake is Windermere and the Southern region has a large industrialized area centered on the Mersey catchment.

The absence of any acute incidents in large lakes in the Northern region provides a striking contrast to the number on the smaller standing water bodies in the southern region.

The problem of acidification of rivers and lakes has recently attracted considerable research and public attention. Effects appear to be confined to Wales, North West England and parts of Scotland, particularly the Galloway region. Fish kills have been reported in some upland reservoirs particularly when new trout stocks have been put in (Swales and Fish 1986). The larger lakes and lowland reservoirs do not appear to be affected but the apparent poor recruitment of trout from tributary streams in Lake Bala is currently under investigation (B. Hodgson pers. comm.).

FACTORS INFLUENCING FISH STOCKS

The climatic conditions existing in Britain have a marked influence on fish recruitment, and strong year classes of coarse fishes are associated with years that have above average temperatures (Mann 1979). This recruitment relationship is also influenced by other factors but as a generalization appears to be true for both rivers and lakes. Detailed long-term data which indicate the importance of temperature are available on population fluctuations of perch and pike in Windermere. Craig and Kipling (1983) and Goldspink (1983) reviewed the factors influencing fish recruitment in the Cheshire meres where temperature is again important for several species. Some of these observations and other studies, for example the Anglian Water Authority's survey data (Linfield 1981), have been used to attempt to inform anglers and to educate them into an understanding that populations do fluctuate naturally. As Mann (1979) succinctly put it:

> ..."The deterioration of your fishery may be the result of man's interference in the environment - but it might not, and you should be prepared to accept the fact."...

Anglers do have a keen perception of the importance of pollution on different systems. The 1970 NOP survey recorded that all fishermen are worried by, and asked for, control of pollution. This was particularly noted for coarse anglers on rivers.

Diseases have had an important impact on fishes, particularly coarse fishes during the last two decades. The two most important angling species, roach and perch (NOP 1970) have both been detrimentally influenced (Bucke et al. 1979). Perch populations have still not recovered in many areas from large scale mortalities that occurred in the late 1960s and early 1970s. These kills were often devastating in their influence. For example, it was estimated that 98 percent of the adult perch population died in Lake Windermere (Bucke et al. 1979). The mortalities could not be associated with any predisposing environmental factors.

SOURCES OF ANGLER AND PUBLIC INFORMATION

The acquisition of knowledge on pollution has many potential sources from the anecdotal to the official. Table 2 illustrates some of the avenues (excluding books and scientific journals) that can be used by an informed lay person, be they angler or concerned member of the public. Perception of environmental problems and a general appreciation of the need to protect the environment have formed the basis for one of the fastest growing participant groups in Britain. Their influence is such that all the major political parties are now including in their policies a commitment to the protection of the environment. The Green Party, although a relatively small grouping, recently won its first local council representation.

It is perhaps worth noting that environmental groups do not necessarily view anglers as protectors of aquatic habitats. Partly this is due to the undoubted confusion caused by the separating of angling from

TABLE 2 SOURCES OF INFORMATION RELATING TO AQUATIC ACIDIFICATION IN BRITAIN

Sources of Information	Information Available
Water Authorities	Data available from public registers.
Welsh Water Authority	Annual report and fact sheet available.
Department of the Environment	Research report outlines acid rain problems and describes funded research etc.
Central Electricity Generating Board	Fact booklets, research booklet and video "Acid Rain" on free loan.
Nature Conservancy Council	Booklet available on acid deposition.
Scottish Development Department	Fact booklet available
Freshwater Biological Association	1984 Annual Report Weekend course in conjunction with Liverpool University
Institute of Fisheries Management	Paper in Annual Study course conference
National Federation of Anglers and National Federation of Specialist Anglers Co-operative Association	Summer 1985 Review pamphlet.
Salmon and Trout Association	Articles by Water Authority personnel in recent magazines.
Friends of the Earth	A great deal of national pamphlets, newspapers, fact sheets etc.
Media	Television eg B.B.C. "Horizon" programme ITV Channel 4 "Worldwise Reports" plus free newspaper available on request etc. National newspapers New Scientist, Angling periodicals.

* This is not an exhaustive list but illustrative of the material available.

conservation and partly it is due to conflicts that have arisen because of discarded lead shot causing mortalities in swans (<u>Cygnus olor</u>) and monofilament line entrapping birds. Anglers for their part have reacted to these accusations and have been at pains to suggest that they were in fact the protectors and often the creators of aquatic habitats while conservation groups are only in their infancy or not yet formed. One hopes that there are now signs that such differences are being resolved and the common ground between the parties are being recognized.

The information sources as outlined in Figure 1 have been divided as far as possible into natural groupings.

Media sources represent a very important category. Both commercial, the Independent Television Authority, (ITA) and National television, the British Broadcasting Corporation (BBC), devote appreciable broadcast time to natural history programs. These add considerably to the general public awareness of the need for conservation both nationally and internationally. Minority viewing groups are catered to by the second T.V. channels, (BBC 2 and ITA Channel 4) and environmental bodies can be provided with their own program slots on ITA Channel 4.

Anglers derive a considerable amount of their knowledge from angling periodicals. Forty-four percent of coarse anglers regularly read the weekly angling magazine and 28 percent regularly read monthly magazines. The respective figures for game anglers was 34 percent and 45 percent (NOP 1980). Examination of these magazines reveals articles, editorials and letters that directly address themselves to pollution problems. It is probable that they are major formulators of opinion in addition to sources of information.

It is perhaps most illustrative to take two case studies of subjects that have gained national importance, to indicate the diversity of information available and to show how opinions may be formulated. The first of these examples is that of acidification. Until relatively recently this was not considered to be a major threat in Britain, although many people were aware of the Scandinavian experience. As an indication of the availability of information the author gathered publically available, published information on the subject. These are documented below together with an indication of media sources, although by no means all of these are shown.

The second example does not relate to a toxin and fishes but a pollutant by-product from anglers: discarded and lost lead weights which have caused swan mortalities. It is not intended to describe the relative merits of the problem here, but these have caused considerable debate. The first indication of

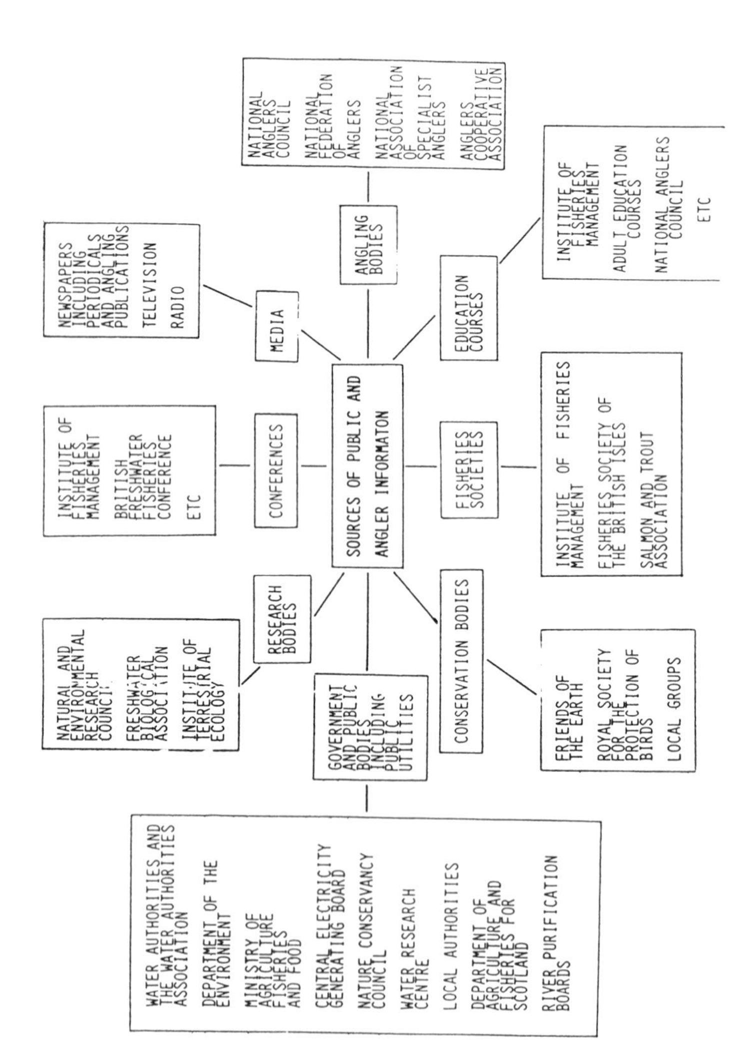

Figure 1. Information sources.

any problem was popularly highlighted in 1981 in the New Scientist (Birkhead 1981), a scientific weekly that is on sale on newstands throughout Britain. From then on the subject attracted considerable attention featuring national TV programs, national and local newspapers, angling newspapers, etc.. The Nature Conservancy Council has issued a booklet on the subject as has the Institute of Fisheries Management (Spillet 1985). Adverse comments on the topic and the threat of the compulsory removal of lead as a source of weights for angling became so severe that manufacturers developed substitutes. Some local authorities have banned the use of lead on their water bodies and the Anglican Water Authority has acted to remove the use of weights above a certain size. The current situation is that legislation will be enacted to ban the use of lead unless the Government is satisfied that anglers are successfully operating a voluntary code.

Similar awareness of concern for pesticides and general habitat degradation could be documented.

SUMMARY

At the time of writing the major directors of fisheries management and water quality control in England and Wales, the Regional Water Authorities, are being considered for privatization and the bodies likely to be responsible for environmental protection have only been outlined in a discussion document. However, it is likely that current practices of fisheries management will remain in the foreseeable future; with most major large coarse fisheries being more or less naturally controlled and trout fisheries in lakes, which are intensively angled, relying on artificial stocking. Pollution control of rivers will remain the principal area of activity, although eutrophication effects will have continued importance for still waters.

ACKNOWLEDGEMENTS

I would like to thank all of the groups and people who have supplied information, often at very short notice. Dr. David Cragg-Hine kindly commented on the paper. The views expressed in this paper are solely my own and not necessarily those of any of the information sources I have used.

REFERENCES

Birkhead, M. 1981. How the fisheries kill the swans. New Sci. 90:14-15.

Bucke, D., Cawley, G.D., Craig, J.F., Pickering, A.D. and Willoughby, L.G. 1979. Further studies of an epizootic of perch Perca fluviatilis L., of uncertain aetiology. J. Fish Dis. 2:297-311.

Chandler, J. 1981. Fishery considerations in recreation: water and land. ed. R.J. Dangerfield, pp. 149-162. London: Institution of Water Engineers and Scientists.

Cooper, M.J. and Wheatley, G.A. 1981. An examination of the fish population in the River Trent, Nottinghamshire using angler catches. . Fish Biol. 19:539-556.

Craig, J.F. and Kipling C. 1983. Reproduction effort versus the environment; case histories of Windermere perch, Perca fluviatilis L., and pike Esox lucius L. J. Fish Biol. 22:713--727.

Edwards. R.W. 1985. Keynote address. J. Fish Biol. 27: (Supplement A):1-7.

Goldspink, C.R. 1983. Observations on the fish populations of the Shropshire-Cheshire meres with particular reference to angling. In: Proc. 2nd Brit. Freshw. Fish. Conf., pp. 251-162. Liverpool University, Liverpool, England.

Graham, T.R. and Pearce, A.S. 1981. Water quality and freshwater fish. In: Proc. 2nd Brit. Freshw. Fish. Conf., pp. 229-239. Liverpool University, Liverpool England.

Hamilton, R.M. 1985. Discharges of pesticides to the rivers Mole and Taw, their accumulation in fish flesh and possible effect on fish stocks. J. Fish Biol. 27 (Supplement A):139-149.

Holland, D.G. and Harding, J.P.C. 1984. Mersey. In: Ecology of European Rivers. ed. B.A. Whitton, pp. 113-144. Oxford: Blackwell Scientific Publications.

LeCren, E.D., Kipling, C. and McCormack, J.C. 1972. Windermere: effects of exploitations and eutrophication on the salmonid community. J. Fish. Res. Bd. Canada. 29;819-832.

Linfield, R.S.J. 1981. The current status of the major coarse fisheries in Anglia. In: Proc. 2nd Brit. Freshw. Fish. Conf. pp. 67-69. Liverpool University, Liverpool, England.

MAFF (Ministry of Agriculture, Fisheries and Food) 1986. Report of the working party on pesticide residues (1982 to 1985). Food Surveillance Paper No. 16. London: H.M.S.O.

Maitland, P.S., Smith, B.D. and Adair, S.M. 1981a. The fish and fisheries. In: The ecology of Scotland's largest lochs. (ed.) P.S. Maitland. pp. 223-251. The Hague: Junk.

Maitland, P.S., Smith, I.R., Bailey-Watts, A.E., Smith, B. D. and Lyle, A.A. 1981b. Comparison and synthesis. In: The ecology of Scotland's largest lochs. ed. P.S. Maitland, pp. 253-283. The Hague: Junk.

Maitland, P.S. 1985. Criteria for the selection of important sites for freshwater fish in the British Isles. Biol. Cons. 31:335-353.

Mann, R.H.K. 1979. Natural fluctuations in fish populations. In: Proc. 1st Brit. Freshw. Fish. Conf. pp. 146-150. Liverpool University, Liverpool, England.

Mason, C.F. 1981. Biology of Freshwater Pollution. London: Longman.

Moore, D.E. 1982. Establishing and maintaining the trout fishery at Rutland Water. Hydrbiol. 88:179-189.

Moss, B. 1983. The Norfolk Broadland: Experiments in the restoration of a complex wetland. Bio. Rev. 58:521-561.

NOP (National Opinion Polls) 1971. National Angling Survey, 1969-1970. London: National Opinion Polls.

NOP (National Opinion Polls) 1980. The National Angling Survey, 1980. London: National Opinion Polls.

Reynolds, C.S. 1979. The limnology of the eutrophic meres of the Shropshire-Cheshire plain: A review. Field Stud. 4:93-173.

Solbe, J.F. de L. G., Cooper, V.A., Willis, C.A. and Mallet, M.J. 1985. Effects of pollutants in freshwaters on European non-salmonid fish. I: Non-metals. J. Fish Biol. 27 (Supplement A):197-207.

Spillet, P.B. 1985. Lead poisoning in swans. Specialist section - recreation. Institute of Fisheries Management. Nottingham, England.

Swales, S. and Fish, J.D. 1986. Angling catch returns as indicators of the status of upland trout lakes. Aquacult. Fish. Mgmt. 17:75-93.

CHAPTER 16

LEGAL AND PUBLIC INFORMATION CHANGE NEEDED

Lee J. Weddig

National Fisheries Institute, Incorporated
Washington, D.C.

It's been almost 20 years since the commercial fishery industry faced its first major crisis involving toxic substances in fishery products. The occasion was the discovery of DDT in Lake Michigan chubs. The latest episode involves the presence of dioxin in Great Lakes fish, a situation currently unfolding. In between these two, the presence of the dieldrin, kepone, PCBs and of course mercury have all created tremendous problems for the United States commercial seafood industry.

Through all of this time, we have yet to see a single documented case of illness caused by the presence of such toxins in commercially caught fish. However, each of the episodes has been disruptive in the market place and has cast aspersions on the safety of our industry's products. Large and small business have incurred tremendous financial losses, not only in the immediate disposal of products judged to be unsafe, but in the availability of supplies and the need to make up lost consumer confidence.

The problem of toxic substances in fishery products is not confined to fresh water species, even though this seems to be the locale of many of the persistent situations. The episode involving kepone originated in a fresh water river but spread into the Chesapeake estuary. PCB presence is noted not only in fresh water lakes but also in coastal areas. Mercury is found in deep ocean fish as well as in fresh water rivers and lakes.

Our prime concern as an industry is the system of law and practice that allows the situation to exist at all. Prevention of degradation and use of our waterways as convenient disposals for chemicals and modern living in general must be our major concern, not only to assure the wholesomeness of fishery products, but also to avoid the negative effects of toxins on the ability of the resources to reproduce and flourish. We have no way of knowing whether the major decline in such stocks as the Atlantic Coast striped bass is due to the fishing pressure, eutrophication, or toxic contamination of waterways where spawning takes place. The matter of preventing further degradation of our waters is a topic for another day. As for now, dealing with the presence of toxins in the water and inevitably in fishery products is part of the way of life for our industry.

In this paper two aspects of dealing with the problem will be commented on: One, the law itself and its interpretation and two, the flow of information to the public.

At the outset it is appropriate to comment on the increasing frustration over the double standard applied to commercial and recreational use of fish. Time after time we have seen how the law prevents the sale of certain products because of toxic presence while the identical product can be consumed by a recreational fisherman's family. The advice given to the sports fisherman is to restrict or to exercise caution in the consumption of the product. Most often the advisory confuses the consumer of a commercial product by implying that that product on the market place should also be restricted.

Part of this dilemma is due to the law itself. Let us take a look at it. The Food, Drug and Cosmetic Act is the basic Federal law which applies to the situation. This law prohibits the sale of food which is adulterated. The law includes a host of definitions of adulteration, but those of pertinence to this discussion are based on section 401 which states that:

> ..."a food shall be deemed to be adulterated if it bears or contains any poisonous or deleterious substances which may render it injurious to health..."

When the substance is not an added substance, such food shall not be considered adulterated under this clause if the quantity of such substance in such food does **"not ordinarily render it injurious to health."** The Act goes on in Section 406 to say that any poisonous substance added to food except where such substance is required in the production thereof or cannot be avoided by good manufacturing practice shall be deemed to be unsafe, unless the Secretary issues regulations, limiting the quantity to the extent that he or she finds necessary for the protection of public health. This section continues by instructing the Secretary that in fixing such tolerances he or she shall take into account the extent to which the use of such substances is required or cannot be avoided. The law's section regulating use of pesticide chemicals on raw agricultural products also figures in the situation in that it prohibits use of such chemicals on raw agricultural products unless a tolerance has been provided. None of these key sections of the Food, Drug and Cosmetic Act was developed with the problem, of toxic presence in fish, in mind. The result is a rather convoluted law which allows the authorities to pick and choose the section they wish when a situation arises.

For example, in the situation involving DDT in Lake Michigan chubs almost 20 years ago, a legal battle was fought over the status of these products, with the authorities maintaining that because DDT was a pesticide without a tolerance for its use on the raw agricultural product of chubs, such were unadulterated. Yet logic demands that we ask what pesticide manufacturer is going to seek a tolerance for residue on fish products in which the pesticide was not directly applied? Great difficulties occur when one considers the basic section 402, especially in relationship to the phases "which may render injurious to health" or, if it is not an added substance, which **"does not ordinarily render it injurious to health."**

This clause of the law has been interpreted to mean that a man-made substance such as PCB would result in adulteration if its presence in fish **"may cause injury"** whereas a natural substance must be shown to **"ordinarily cause injury."**

A rather peculiar situation came about in the early years following the discovery of mercury in fishery products. Mercury is a natural element in the environment as evidenced by its presence in early museum samples of fish preserved well before heavy industrialization. However, the FDA took the attitude that mercury was an added substance and therefore, set its action level on the basis that "may cause" injury. This case was taken to court. The judge in Solomon-like fashion ruled that part of the mercury in the swordfish was human-made, while another part was natural. Therefore, no one won. The end result, however, was that the action level for mercury was raised from 0.5 to 1.0 part per million. The point to be made is that the environment is not static; it is hardly pristine after these many tens of thousands of years of human existence. Human presence on the planet has altered the environment so that in applying the Food, Drug and Cosmetic Act to fishery products, the concept of trying to differ between natural and man-made is no longer logical. The natural environment as it now exists does include toxic chemicals, like it or not. There is no way of determining the specific source of PCB, mercury or dieldrin in the environment. Whether toxin is man-made or the result of volcanic action does not alter its toxicity.

Another problem concerns the procedure the Food and Drug Administration (FDA) has been using to determine action levels. The FDA has only rarely used the tolerance setting procedure in handling the problem areas. Rather, it has utilized action levels or guidelines as the way to determine when a product could or could not be taken to the market. FDA's ability to utilize the action level approach as opposed to the formal tolerance setting regulatory process has been challenged and was argued before the Supreme Court in May of 1986. For the commercial seafood industry, the present approach must be upheld. If the challenger's point of view prevails, it would mean that the presence of any measurable toxic substance within seafood would automatically cause adulteration unless a tolerance were set. The tolerance setting procedure is extremely time consuming and expensive. The industry would be out of business entirely in view of the newer analytical techniques which allow measurement of foreign substances at levels of part per trillion. To place the matter in perspective, setting a tolerance for PCB took seven years!

All of these problems with the Food, Drug and Cosmetic Act dictate need for a new section which would apply specifically to the problem of the environmental contaminants in fish and seafood products. The new section should recognize that the toxic substances are not added in the sense of additives used in the processing procedure or pesticides used in agricultural contents. The section should provide for a rapid method of determining action or regulatory levels so as to avoid lengthy procedures.

The present methodology of determining action levels itself is flawed in several respects. The Food and

Drug Administration interprets section 406 of the law to mean a drop in the residues is reason to reduce the action level. The industry never gets relief as the fundamental problems of the contamination are alleviated through time. Our feeling is that since the substance cannot be avoided at all, there is no justification in reducing actions levels as soon as levels begin to drop. A further flaw in the process concerns analysis of economic loss. Typically, the regulators will calculate that if a certain percentage of a total product mix is in excess of a given action level then the economic loss will be the removal of that percentage of the product from the market place. This thought process is faulty in that it does not recognize the impossibility of separating the above guideline product from that which is below the prescribed level. Losses most often will be the entire production. There are exceptions when the product can be segregated by size and location of catch, but this situation does not occur very often.

Difficulties with the present action level system continue after the level is determined. Assume a tolerance of one part per million. Since it is impossible to test every fish, a sampling procedure must be followed. Samples are drawn, a composite is made and analyzed. If the composite is over 1 part per million, the lot is considered unsafe and rejected. If the test results show less than 1 part per million, it is considered safe and can move into the market place. Yet, when the producer ships the products, another inspection agency may pull one fish or another at a point further along the distribution chain. This particular fish may very well be over the guideline with resultant clamour about unsafe fish on the market place, with attendant expense. In short, the producer is in double jeopardy.

The entire attitude about the meaning of guidelines or tolerances is without logic. There is a feeling that health is affected by a single exposure to an above-guideline product. In the earlier days of the mercury situation, the understanding was so poor that TV stations would film testing situations, going from one fish to another: This one is good, this one is bad. The bad ones had mercury content slightly over 0.5 ppm while the so-called good ones would be slightly under 0.5 ppm. There was no understanding that the presence of such toxin has impact only through long-term exposure and that the concept is to maintain average levels below a certain point.

A more intelligent use of a guideline would be to apply it at the production level only and eliminate entirely testing further in the marketplace. If, on average, products from a certain body of water are under the guideline, they should be allowed to move freely. If on average, they are above the guideline, then the production area should be closed. The concept could apply to the entire production from a body of water, or to a class of products, such as those larger than a certain size or taken in a certain time of the year or in a certain location.

Still another difficulty concerns the immediate reaction of the authorities after a potential problem is identified. Assume a toxic substance is discovered. An appropriate action level is set. Immediately, the reaction is to clear all shelves and inventory of product to prevent a single additional ingestion of the food. This may be logical if the single substance has a possibility in its own right of causing harm. Since most of these situations involve potential harm due to long-term exposure, it does not make sense to be overly concerned about using food already in the distribution chain despite the fact that it may contain quantities of offending substances slightly above a newly formed action level.

The author vividly recalls the need to destroy thousands of pounds of swordfish at the time mercury was discovered to be harmful, just because it contained more than 0.5 ppm of mercury. Products containing slightly under 0.5 ppm was deemed to be okay. Looking back at the episode now it seems illogical that any such product was destroyed when potential harm could occur only through a long-term exposure.

These problems with enforcement of action levels lead us to a consideration of the second part of our discussion, the need to improve the information flow to the consumer. Unfortunately, the means of communication with the consumer do not lend themselves to a measured, logical information flow. The meaning of tolerances or action levels, the relationship of such to overall diet, exposure and general health, all are complicated subjects. The elements of news reporting, especially in television, do not lend themselves to education. The stories relating to mercury began with pictures of deformed people in Japan and led quickly to prospects of the same in the United States due to consumption of fishery products found to contain mercury here. The fact that the situation was entirely different, and the levels found in the product were far different, was not considered to be newsworthy. The end result was hysteria.

Similarly, the present structure in which advisory opinions are issued by state regulatory agencies lends itself to tremendous confusion in the marketplace. The difficulty of applying one law to the commercial harvest and another to recreational use of the same item was identified earlier. A good example of how distribution of scientific information and advisories can go astray occurred just as this past week with the release of somewhat preliminary findings by the EPA from a survey of the level of dioxin in the Great Lakes region.

The science involved the study of dioxin presence, mainly in harbour dwelling carp and the fillets of large trout caught in the open waters. Dioxin presence was determined, ranging from 0 to about 20 parts per trillion. There is no formal action level from the FDA on this substance, although the FDA has informally expressed concern when products exceed 25-50 parts per trillion. Several state officials have a more rigid attitude on potential harm.

Although the EPA study did not involve fish other than carp and large trout, extrapolations were made that dioxin could also be present in other fish found in the Lakes. The result of all this was a headline: "Cancer Link Dioxin Found in Lake Michigan Fish" with a box that says which fish to avoid. The box went on to say that no one should eat a number of fish including walleye larger than 20 inches, brown trout, carp, lake trout larger than 25 inches, white sucker, catfish and white bass. It gave further restrictions, including large chinook salmon for women and children. It showed a category of fish posing the lowest risk, including smelt, perch and coho salmon. The problem with this story is that while it may provide appropriate advice for people who are out catching fish from Lake Michigan it implies that all walleye pike larger than 20 inches should be avoided, that large lake trout should be avoided, that catfish should be avoided. It also suggests that women and children should not eat chinook salmon larger than 25 inches.

From a commercial point of view, this is extremely distressing since the catfish on the market are farm-raised from controlled aquaculture operations in the south. Lake trout on the market most often comes from Canada, while chinook salmon would be coming from the North Pacific. Walleye pike most often would be coming from Canadian lakes since most of the U.S. states prohibit the catch of walleye for commercial purposes. All rainbow trout on the commercial market come from trout farms and mountain states.

In short, the advisory to sports fishermen has impacted the commercial status of many species even though they are not involved whatsoever with the problem. In this particular instance the story was picked-up by the media in Europe. The result was inquiries from the buyers in Europe who were raising questions as to the safety of all fish from the U.S. and Canada. This is not hard to understand since most of the U.S. exports to Europe are salmon caught in the Pacific Northwest. The report said women and children should not eat chinook salmon larger than 25 inches, with out defining this as being sports-caught fish from Lake Michigan. It becomes garbled and creates an international stir. For a while, it appeared that the Swiss government might ban all fish from the U.S. and Canada because of this flawed news reporting.

In looking over the report and fact sheet issued by EPA, it is difficulty to be overly critical in that the subject was treated in straight forward fashion. Yet, nowhere does the information cover the relationship to commercially caught product. One might say that this is not the job of the EPA. Yet the nation's interests are not served by generating international incidents about the quality of the seafood products moving into Europe. Perhaps, a partial solution to this problem might occur if advance advice were given to people who could explain to and help the press understand the overall perspective of a story. When the government agency concerned with fishery products does not know of the release of information that has a damaging effect on the seafood industry, nor when trade associations such as National Fisheries Institute reads about new information when it has already made headlines in this fashion, it is too late to be of much assistance in making certain that the story does appear in the proper perspective.

Is anything intrinsically wrong with sharing new information, prior to its release to the public, with all of the factions that could be affected by it?

It also makes sense to have a uniform determination of advisories throughout the nation as opposed to the present set-up where different health authorities can and do have contrary attitudes on the implications of various new findings.

In conclusion, we can summarize that in our opinion, present law needs modification in order to adapt to, and more effectively regulate the growing concern with toxic pollutants in fisheries resources. The methods of releasing information in order to effectively advise the consumer of risks also needs improvement.

While we have what we believe are legitimate criticisms of the current system, we do not want to suggest in any fashion that potential health risks are to be ignored or covered up. The long-term well-being of our industry depends on maintaining a supply of safe and wholesome products. We also recognize that safety and wholesomeness are not necessarily clear cut when dealing with long-term effects of chemicals for which prior experience or understanding is not present.

We do suggest that as the regulatory agencies look at their activities, they consider the impact on the commercial industry and recognize that the industry is not a monolithic, impersonal entity, but rather is made up of individual citizens trying to earn a living in a rather difficult business. Impacts on the industry can, of necessity, be measured by dollars and cents lost and analysis of food removed from the marketplace. The true impact though is the disruption in the lives of working men and women and their families, as their businesses fail, investments are lost and ways of life are destroyed.

CHAPTER 17

PREDICTING, VALIDATING AND MONITORING EFFECTS OF TOXICS ON LARGE LAKE ECOSYSTEMS

John Cairns, Jr.

University Center for Environmental Studies, and Department of Biology, Virginia Polytechnic Institute and State University, Blacksburg, Virginia 24061 USA

ABSTRACT

Most information on toxics has been generated with single species laboratory tests. Many species used are not key species in large lakes and the test conditions are low in environmental realism. Validation of predictions in rivers and small lakes is rare and almost non-existent in large lake ecosystem conditions. Similarly, most monitoring protocols and methods are designed to be used in streams, rivers, reservoirs, or small lakes. The lack of available information introduces an uncertainty into management decisions for large lakes and such decisions are rarely as defensible as managers (and the public) would like them to be. A strategy for improving this situation is given in this discussion.

INTRODUCTION

Gathering information and improving decisions should be based on that information can be augmented by an array of tests at different levels of biological organization. The tests should be carried out in a sequence of tiers so that information is generated in an orderly, systematic fashion. These are:

Tier 1 - screening or range finding tests;
Tier 2 - predictive tests;
Tier 3 - validating tests, and
Tier 4 - monitoring.

This will strengthen the present system that is deficient in several important aspects:

1. little or no attention to levels of biological organization above single species,

2. little or no validation of predictions, and

3. insufficient monitoring.

WHAT CONFIDENCE CAN WE HAVE IN ESTIMATES OF TOXIC EFFECTS?

If we were all to stand at the Straits of Mackinac between Lakes Michigan and Huron with a standard single species toxicity test unit for fathead minnows or *Daphnia* at our side and ask:

> ..."Do we really believe that we can predict toxicological events in these two large lakes from dose-response data generated in these small containers?"...

The author wonders how many participants in this conference would be willing to say **"yes"** and stake their scientific reputations upon that judgment. The conveners of this symposium chose the location wisely not because of the amenities, although these are indeed splendid, but because we either arrived by boat or by plane and moved over a part of the Great Lakes system in doing so. The difference in ecological complexity between the simple single species test system (and the author includes here the most elaborate and sophisticated of these) and any one of the world's great lakes is abundantly clear. But, perhaps, our knowledge of the ecological function of the world's large lakes is so good that given some basic toxicological information we can extrapolate from the responses of single species (such as lethality, behavior, reproductive success, respiration and the like) to critical functional attributes of large lakes, such as nutrient cycling and energy flow. Given our present predictive capability, this notion can only be charitably described as preposterous.

There is another possibility: by selecting the most sensitive species from the system itself or by selecting a test species more sensitive than any of the indigenous species, we can ensure the protection of all species. A corollary assumption is that if all species are protected individually, their aggregate (i.e. community) function will also be protected. An earlier, briefer examination of this question may be found in Cairns (in press). However, in the interest of completeness of this discussion and because of the importance of the topic, a somewhat extended discussion of this critical issue follows.

THE MOST SENSITIVE SPECIES

For the last thirty years, aquatic toxicologists have been engrossed in the search for the **"most sensitive species."** If one finds this most sensitive species and sets standards for presence of toxic materials and other stressors in natural systems accordingly, presumably all other species and all other activities at higher levels of biological organization will be protected. A recent statement of this assumption has been made by Weis (1985) commenting on an article by Kimball and Levin (1984) espousing the need for tests at higher levels of biological organization than single species.

A number of assumptions are inherent in the most sensitive species approach, although these are not always explicitly stated:

1. The selection of most sensitive species from an array of customarily used test organisms means that there is some correspondence with the responses of the most sensitive species in a much larger array of exposed organisms in natural systems.

2. There are no significant responses at any level of biological organization that are more sensitive than the end points chosen for the most sensitive species toxicity tests.

3. The savings resulting from using the **"most sensitive species approach"** are not offset by the costs of making bad management decisions.

4. A species shown to be **"most sensitive"** to a limited array of toxic substances will invariably be so for a much larger array of

toxic substances. This is a particularly important point, because once the most sensitive species has been selected, repeated validation and confirmation of this assumption may not be as rigorous as one would wish.

Assumption 1

The selection of the most sensitive species from an array of customarily used test organisms means that there is some correspondence with the responses of the most sensitive species in a much larger array of exposed organisms in natural systems. The very commonly used test organisms for aquatic single species toxicity tests probably do not exceed 20 for the contiguous United States. Add those less commonly used to the list with the requirement that they be used by at least five investigators with reasonable regularity, and the list would probably not exceed 50 species.

For the purposes of this discussion, however, assume that the list of test species in reasonably common use is 100. A recent field investigation of the Flint River carried out primarily by the Academy of Natural Sciences of Philadelphia with some assistance from Florida State University and Virginia Polytechnic Institute and State University identified almost 1000 species in the Flint River and Lake Blackshear. The study area involved <10% of the total drainage of the main stream of the Flint River. The research covered a time span of <2 years and, therefore, did not fully display successional processes by any means. Although this was not the intent of the survey, the species list would have at least doubled had it been carried over for a 10 or 20 year span of time. If the species list were extended to aquatic insect larvae only for all freshwater aquatic systems in the contiguous United States, many thousands of species would be observed. Therefore, the probability of actually using the most sensitive species, not on the basis of a list of species commonly used in laboratory toxicity tests but rather for the entire array of aquatic ecosystems in the contiguous United States, seems rather remote. If it is the most sensitive species only for the limited array of commonly used test species (assuming for the moment that it is so for all chemicals in all test conditions), what does this mean in terms of management and regulatory decisions? The answer is, of course, that sound judgments in these areas will only be possible if one can extrapolate or predict, based on the results from the most sensitive species tests, the responses of a larger array of species and at different levels of biological organization. Since the most sensitive species was not selected for this purpose, it seems highly probable that these extrapolations and predictions will be scientifically justifiable and will live up to expectations of the users and the general public.

If the **"most sensitive species"** should actually be vastly more sensitive than the other species in the natural system, one might reasonably ask if the expenditure of substantial amounts of money for its protection is justified. Keep in mind that this is a social-political, not a scientific decision, although scientists might want to attempt to persuade those making the decision that the total protection thus afforded is desirable. The decision is more difficult if the most sensitive species is not indigenous to the natural system in question - a situation that is not infrequent where a single test species is highly touted for toxicity testing in the entire contiguous United States including the Great Lakes. One is then providing a degree of protection that would be essential were the species indigenous, but since it is not, one might reasonably question the use of this level of protection.

Assumption 2

There are not significant responses at any level of biological organization that are more sensitive than the end points chosen for the most sensitive species toxicity test. Although this assumption is rarely explicitly stated, it is a commonly held belief, as most recently evidenced by a direct statement to this effect:

> ..."It seems to me, therefore, that in looking for the effects of toxicants, single species tests may be the most sensitive if one chooses the species and parameters carefully"...(Weis 1985).

In its simplest form, this assumption should read: There is no end-point at any level of biological organization that is more sensitive to toxicant or other stress than the end points used for the most sensitive species toxicity test. In short, if there is no-observable response in the **"most sensitive species"** toxicity test, it is inconceivable that any other level of biological organization, including community and ecosystem could be adversely affected. It is important to determine whether those who believe this, base their faith on presently used end-points or all the possible end-points that might be used with the most sensitive species. If the intent is to use all possible end-points for the most sensitive species tests, even though they are not presently used, the statement may have theoretical value as a hypothesis. However, in the absence of supporting evidence, the statement has no present practical value. If, on the other hand, the growth is based on presently used end points (such as lethality, growth, reproductive success, nutrient uptake and the like) for the sensitive species toxicity tests, one can analyze the probability of the soundness of this hypothesis. The hypothesis would now read:

> ..."If no deleterious effects are observed for the end-points presently used in the most sensitive species toxicity tests, no deleterious effects will occur in aquatic communities or ecosystems (even if there are adverse effects on the most sensitive species in characteristics not presently being used as end-points)."...

One must then ask whether the end-points presently used in most sensitive species toxicity tests for all chemicals and all test conditions, are invariable the most sensitive of the total array that one might possibly use. It is difficulty to imagine a prudent scientist answering this question in the affirmative.

End-points for toxicity tests are often chosen for reasons of convenience and replicability rather than for sensitivity. For example, Doane et al. (1984) showed that changes in respiratory rate were appreciably more sensitive to a number of toxicants than a variety of commonly used end-points including blood chemistry, tissue damage, etc. However, respiratory rhythm is not commonly used in most sensitive species toxicity tests. Some other end-points commonly used in those tests are shown by Doane et al. (1984) to be less sensitive, at least to the chemicals tested. The publication just cited made no claim to using the most sensitive species but it is worth noting the relative sensitivity of the end-points chosen. If one assumes a high probability that there are end-points not presently used that are more sensitive for some chemicals and some test conditions than the end-points presently in use, one can assume that the **"no observable-deleterious-effects"** for certain end-points do not guarantee that all end-points will be unaffected. One is then forced into the use of application factors, as is the case for all test species. If these are reliable, it is difficult to justify the predominant use of the most sensitive species.

If one is forced to use a large array of end-points in most sensitive species toxicity tests (assuming for the moment that there is no end point at a higher level of biological organization more sensitive than one of these end-points), it is likely to increase the cost substantially, even if the assumption is valid. In our laboratory (Niederlehner et al. 1985), some community level toxicity tests were recently carried out with protozoans. Briefly the test organisms were collected on an artificial substrate and brought into the laboratory where they were placed in protozoan-free water with a series of uncolonized artificial substrates. A concentration of cadmium that did not appear to injure the species on the source pool (i.e. the substrate colonized in a natural system) did markedly reduced the colonization rates when compared to a control unit without cadmium. One could postulate that the examination was not sufficiently thorough to detect subtle changes in a large array of species - a charge that is absolutely correct. However, this is just the point. Without knowing that the colonization rate had been affected, would one have looked for more subtle effects on the source pool community? The most likely explanation for the fact that the species on the source pool remained unaffected is that micro-habitat conditions there protected them in some way (e.g. by tying by the cadmium), but this protection was lost when the organisms left the colonized substrate and were en route to the new uncolonized substrate. The control unit without cadmium had a colonization rate that was normal for such test systems.

The major consideration, however, is that even a modest degree of environmental realism produced a response at a concentration of cadmium that would ordinarily have been presumed **"safe"** or non-deleterious to the array of species present. Notice that it was not the species that changed but rather than an activity was involved in this particular community level test that would not normally be present in a toxicity test, although it is routine in natural systems. Engaging in this particular activity increased the vulnerability of species that were otherwise presumably unaffected.

In addition to this example, other evidence that multispecies toxicity tests furnish information useful in making regulatory and management decisions may be found in Cairns (1985, 1986).

One might postulate that even colonization rate or successional processes that depend on new colonists could be accurately predicted if the right single species were used in a test such as the one described. While this may be true, the author does not find the argument persuasive, given the present state of knowledge about the colonization process and succession (see the discussion by McIntosh 1980).

If the basic assumption concerning no significant responses at any level of biological organization that are more sensitive than the end-points chosen for the most sensitive species toxicity tests is correct, or substantially correct, evidence should be provided to validate this assumption. Even if this hypothesis is validated, it would not in itself justify using the most sensitive species approach as the following discussion shows.

Assumption 3

The savings resulting from using the **"most sensitive species approach"** are not offset by the cost of making bad management decisions.

Even if Assumptions 1 and 2 are found to be essentially correct, the most sensitive species strategy may not prove useful from either a regulatory or an industrial management standpoint. It is not at all unlikely that if Assumptions 1 and 2 are correct, they will be so because no other species or aggregation of species even approaches this one in sensitivity to toxicant and other stress under a variety of test conditions. It seems highly improbable that Assumptions 1 and 2 will be confirmed for a species that is only slightly more sensitive than all other species and aggregations of species, but rather that such a species will be markedly more sensitive. To state the situation differently, this species could be dramatically affected by concentrations of chemicals that would have no-observable effects on either the structure or function of most other species or aggregations of species. Regulatory agencies would then be requiring substantial waste treatment costs for industries and municipalities for the protection of a single species that is probably not resident in most of the ecosystems receiving the potential contaminant.

This is not an unlikely scenario since a very wide geographical and ecological distribution hardly seems compatible with ultrasensitivity to stresses - even those of anthropogenic origin. It is no accident that some of the most commonly used test species, such as the rainbow trout and the bluegill sunfish, are widely used in stocking programs because of their adaptability. One wonders how well a very sensitive species could be kept alive in the laboratory for routine use. Safety factors are, of course, as essential for ecological systems as they are in elevators, bridges and automobiles. However, the determination of the degree of safety is not a scientific judgment, whereas the probability of harm or risk is.

As a consequence, it seems appropriate to distinguish these two activities and not attempt to incorporate them into a single test. If we are to make sound, regulatory and management decisions on environmental protection, it is important that we know the response of complex systems to a variety of concentrations of chemicals and other stresses along a response gradient. Having established this gradient, we will then make a decision on the desired level of protection (i.e. end-points or characteristics that we wish to be entirely unaffected) and, finally, may add some safety factors to allow for uncertainties about the operation of the waste treatment system, the fate and transformation of the chemical in natural systems, and inaccuracies in estimating the response of complex systems from laboratory tests. We may even wish to install a biological monitoring system accompanied by selected chemical and physical monitoring to verify that our assumption of protection for selected characteristics is sound. The most sensitive species approach attempts to bypass or ignore this complex array of decisions and information, it may be one of the most expensive tests possible even though the costs are not incurred in carrying out the test itself.

Assumption 4

A species shown to be **"most sensitive"** to a limited array of toxic substances will invariably be so for a much larger array of toxic substances. The sea lamprey in the Great Lakes is not usually thought to be a species particularly sensitive to various chemicals. Yet, after a search that involved testing literally thousands of different chemicals, one particular chemical was found (TFM) to which the lamprey was significantly more sensitive than a variety of **"desirable species"**, many of which were considered to be relatively sensitive (e.g. the rainbow trout) to a variety of chemicals. The hypothesis of the U.S. Fish and Wildlife Service, which ultimately proved to be correct, was that a chemical existed to which the lamprey was significantly more sensitive than were other species inhabiting the same ecosystem. The difference in sensitivity turned out not to be substantive, but it was real.

Of course, not all of the thousands of species in the Great Lakes associated with the lamprey have been studied extensively, so some may be even more sensitive to TFM than the lamprey. Presumable if sufficient resources were available, almost any species could be selected for carrying out the same exercise with equal effectiveness. In short, relative sensitivity to Chemical X for two or more species is probably not going to be true for Chemical Y. Therefore, the degree of sensitivity of the most sensitive species will probably vary enormously from test to test.

This situation does not pose an insurmountable problem if one is using an array of evidence, particularly at different levels of biological organization, but might result in major errors in judgment if the most sensitive species test is the sole source of information. The most sensitive species may only exhibit this characteristic in a natural environment or a laboratory test situation with high environmental realism. Conversely, a species that is extremely sensitive in the laboratory may be only so because its environmental requirements are not being met.

In many of his articles, the author has affirmed his belief in properly used single species toxicity tests (e.g. Cairns 1983). They are now, and probably will continue to be for some time, the backbone of our efforts to determine the probability of harm to more complex systems. The author is deeply disturbed when people use simple laboratory tests of short duration to infer a degree of protection for the environment that is not warranted by the evidence at hand.

Multispecies toxicity tests including micro- and mesocosms are not now commonly used but they do provide information not usually generated in single species tests. Prudently used and interpreted, this additional evidence should further enhance our ability to protect natural systems from the wastes and products of an advanced technological society. To retreat to the single most sensitive species toxicity test as espoused in the letter by Weis (1985) is to ignore all the knowledge of the complexity of ecosystems that has been painfully accumulated over the last 30 or more years.

A somewhat modified version of the most sensitive species concept is that of keystone or controlling species. These influence associated species in ecological communities to a much greater extent than their abundance or biomass would indicate (e.g. Janzen 1966, Paine 1974). The assumption in this case is that if keystone species are identified in terms of regulation, control of energy flow, and the like, are used in the toxicity tests, then the information will be adequate to protect the entire system. Anyone examining the functional flow diagrams for large lakes realizes the difficulty of identifying all keystone species and also realizes that if substantial numbers of non-keystone species are affected then the keystone species would almost certainly be affected as well.

Another widely used alternative to the direct measurement of response to complex systems is the application factor (e.g. Petrocelli 1985). A number is used as a multiplier of results gained from a toxicity test to offset some of the deficiencies of the test itself. Although these deficiencies are rarely stated explicitly in conjunction with a description of the development of an application factor, one might assume them to be identifiable as general deficiencies resulting from low environmental realism or, alternatively, the differences between the test system and the natural system.

A partial list of these deficiencies follows:

1. Sensitivity in the test species is not adequately represented in the test population.

2. Changes in water quality (e.g. hardness, pH, dissolved oxygen concentration) that would mediate toxicity do not replicate those in natural systems.

3. Responses of other life history stages of the test species often are not included in the tests.

4. Responses of most other species inhabiting the receiving system regularly are not included in the tests.

5. Responses of other levels of biological organization (e.g. community, ecosystem) are not included in the tests. For example, a single species toxicity test application factor should include responses at community

and ecosystem levels, if such protection is implied.

6. Interactions with other chemicals that might make the response additive or synergistic are rarely extensively studied. Cumulative impacts do not receive the attention they deserve (National Research Council 1986).

7. Effects of condition factors (organismal) are not included in the tests. For example, if practically no control mortality is mandatory, the animals are likely to be in better condition than they are in many natural systems. However, poor condition might produce more susceptibility. Control organisms are often parasite free or may have had parasitism reduced by prophylactic treatment that would make the organisms more resistant to a toxicant than they might otherwise be.

8. A longer period of exposure often occurs in natural systems than was possible in the test. Few tests are carried out for the lifetime of vertebrates, or even for many macrointertebrates. Tests carried out for several generations are exceedingly rare. However, if exposure might be for a long period of time, extrapolation to the no-response threshold following long exposure must be made.

9. Deficiencies caused by problems of scale. The test container may alter the response because it is too small or in some way different from a natural system. This is particularly important to large lakes.

10. Deficiencies caused by lack of environmental realism in the test container or device; also very important for large lakes.

11. The margin of safety the public or their representatives consider desirable should be included.

Gross, highly visible damage is probably prevented by the use of single species toxicity tests in combination with a fairly substantial application factor (i.e. one resulting in an acceptable concentration that is much lower than the response concentration in the test itself). The author has some reservations about both the direct economic costs of following this procedure as well as the ecological cost.

A second and highly probable reason for lack of visible problems is the high degree of variability of natural systems coupled with cyclic phenomena (e.g. periodic floods, changes in water level in large lakes, etc.). Odum et al. (1979) discussed this variability and hypothesized that as stress on a complex ecological system increases one of the first changes may well be increased variability. Since natural variability of most complex systems is poorly documented, it would be exceedingly difficult to detect a 10% or 20% increase and even a 50% increase might go unremarked if there were not a substantial historic data base and the system were not continuously monitored.

A fine example of the type of argument on subtle, and not so subtle, ecosystem changes has been occurring in the acid rain effects debate. Acid rain covers such large geographic areas that gradients can be established and so on. Suppose that the phenomena were restricted to one large lake instead of a number of small and large lakes over a wide geographic area. The amount of evidence deemed conclusive would probably increase by an order of magnitude on a particular system.

Finally, a generalized application factor designed to include every probable deficiency of a toxicity test will probably end up being a worst possible case number most of the time. Although, theoretically, all deficiencies might be important for a particular ecosystem, it is more likely that some will be important, some will be unimportant or negligible, and some will be moderately important. Even if some important deficiencies are omitted in the development of the application factor, these will be offset by some that are included but are unimportant for a particular ecosystem or situation. As a consequence, the very lack of precision in the ability to extrapolate from single species to large lake ecosystem effects and recognition of it in the development of the application factor may be the single most important factor in the ability of single species toxicity tests to work well. However, it is worth emphasizing that, if this is true, running the toxicity tests may not be particularly costly, but meeting the conditions dictated by the tests plus the application factor in terms of waste treatment costs, and so on may be very great indeed.

DEVELOPING A TOXICS ASSESSMENT STRATEGY

The executive summary of a report by the Committee to Review Methods for Ecotoxicology of the Commission on Natural Resources of the National Research Council (1981) states:

> ..."Single-species tests, if appropriately conducted, have a place in evaluating a number of phenomena affecting an ecosystem. However, they would be of greatest value if used in combination with tests that can provide data on population interactions and ecosystem processes."...

The executive summary also states that the report should not be interpreted as a criticism of single species testing, but the report is critical, however, of using single species data to predict effects of chemicals upon interactions within and among species (e.g. competition, predation and relationships between hosts and parasites) and upon effects at the system level (e.g. alterations in flow of energy, nutrient spiraling and diversity). Cairns (1983a) acknowledges that single species tests have been **"the workhorse"** of the aquatic toxicity testing field. However, the National Research Council report (1981), Cairns (1980 and 1981), Kimball and Levin (1985) and others question the advisability of using single species tests to predict community and ecosystem responses.

The means to resolve the issue of determining the effects of toxics on aquatic ecosystems have been available for years. The first of the Pellston Workshop Series (Cairns et al. 1978), held only a few miles from the Large Lakes Symposium site, espouses the orderly and systematic acquisition of data on both environmental concentration of a chemical and its environmental effects in sequential series or testing tiers. The second Pellston Workshop (Dickson et al. 1979) provided some illustrative protocols from industrialized countries, with more explicit directions on the information that should be gathered in the various tiers. Cairns (1980-81) noted that the usual sequential arrangement of tests from the simple to the more complex possibility reflects in a broad general way the historic development of the field. Therefore, ecotoxicologists placed toxicity tests, with which they had long familiarity, early in the sequence and placed last the more recent and more sophisticated tests that are still in the experimental stage of development.

As an alternative, Cairns (1983b) suggests simultaneous testing at all levels of biological organization to determine which are the most critical thresholds for a particular decision. This simultaneous generation of information, focused on a specific decision, would soon identify which information is redundant and which is not. An alternative safeguard is the process of validation or confirmation. If the predictions made from the evidence gathered in Tier 1 and Tier 2 about community and ecosystem responses (or lack thereof) are confirmed or validated (Tier 3) in a scientifically justifiable way, the errors inherent in this process should be immediately apparent.

The validation process (this word is now being used more frequently than confirmation) should identify when there is a high correspondence between the predictions based on laboratory evidence and the response in natural systems (e.g. Livingston et al. 1985). For those areas where the correspondence is high, no further evidence need be gathered, and additional tests would probably not be a good investment, at least for regulatory purposes. If ecosystem responses occur at concentrations of toxics where no responses were predicted, clearly some alternative tests or procedures are mandatory. Continuous monitoring (Tier 4), to ensure that quality control conditions and ecosystem integrity, are being preserved will provide a backup judgment of the long-term efficacy of the validation process and will simultaneously provide additional evidence on the normal variability of the natural systems.

There are a number of levels at which the validation or confirmation process could be carried out (Sanders 1985, Cairns in press b, c, d). Krebs and Burns (1977) had an interesting field and laboratory comparison of the fiddler crab (<u>Uca pugnax</u>). A reverse case where laboratory experiments supported field observations was carried out by Svedmark et al. (1971). Dicks (1973) compared laboratory and field observations on the limpet. Cairns and Cherry (1983) found there was good correspondence between single species laboratory toxicity tests on fish and invertebrates and the same species in a natural system.

There are a number of reasons why this correspondence was probably so high. Among these, the most important may have been:

1. the laboratory was right beside the river being studied and water from the river was pumped directly to the laboratory with little

delay,

2. time for transport of test specimen from field to laboratory was minimal, and

3. indigenous species were used for both field and laboratory studies.

A high correspondence might not be expected for field and laboratory studies in which the laboratory species was not an indigenous species and the species used for field validation was. The correspondence might be particularly poor when a non-indigenous test species commonly found in rivers or small ponds and lakes is used to make predictions about the response of indigenous species characteristic of large lakes.

One wonders if the laboratory test species would tolerate or function well under test conditions environmentally similar to those of the large lake and, if not, clearly the results would be distorted. If conditions more suitable to the test species than representative of the large lake were used, one wonders about extrapolations of results from a dissimilar system to a large lake. Some parallel concerns on chemical fate models may be found in Herbes (1986).

Of course, extrapolating from one level of biological organization to another is probably at least an order of magnitude more difficult than from a species in the laboratory to the same species in the field or even from one species in the laboratory to another species in the field. Although present evidence is not adequate to document this, the number of variables affecting a more complex system such as a community are certainly going to be much larger than those affecting a single species. The interactions of each species with other species in the community together with the environment will certainly be considerably greater than those of a single species with its environment.

Finally, ecotoxicologists must incorporate system attributes that are not even visible at the lower level of biological organization. A concrete example involving large lakes would be to test hypothetical Chemical X using life cycle end-points for cladocerans or fathead minnow (e.g. Mount and Norberg 1984, Norberg and Mount 1985) and extrapolate from these results to the ability of a phytoplankton community to fix carbon in a large lake system. Of course, the author has never seen an explicit statement that this is the intent of a single species test. On the other hand, neither has the author seen a warning that the test is not suitable for such extrapolations.

The challenge the author makes at this conference is this:

> ..."if people believe all of this can be done with single species tests (and it may be possible), then the problem is too important not to have scientific evidence directly supporting this assumption. If the evidence shows it is not possible or those who defend the use of single species tests think it is not a reasonable extrapolation to make, then it is high time to begin working on how to protect important biological functions above the level of single species responses."...

As an intellectual exercise, consider another important attribute of large lake ecosystems - namely the ability of macroinvertebrates to harvest phytoplankton, or, at another level of biological organization, the ability of the diatom community to utilize silicon. In most large lakes, diatoms are important both in the water column and attached to various substrates. Presumably the ability to utilize silicon is important to them, but, if ecotoxicologists are uncomfortable with the most sensitive species assumption or the assumptions involved in the use of application factors, then there is a serious credibility problem about our predictions of toxic effects in large lake ecosystems.

CONCLUDING REMARKS

The author realizes that at least some of the objections will be raised by representatives of regulatory agencies, industry and even members of the academic community. Regulatory personnel will claim that multispecies tests, microcosms, mesocosms, and the like are difficult to run, are unproven, are more costly, require more professional training and a variety of other objections.

The principal objection will be that the present system is working well, why question it. Not many months ago we saw what happened with the Challenger Space Shuttle, which had worked well for a large number of flights and suddenly failed. Highly complex systems, such as space shuttles or ecosystems, warrant our suspicions of protective measures and demand continual proof of their efficacy. As in the Challenger incident, it may not be so much that the system is working well but rather that there has been, up until now, no dramatic recorded failures. In the case of toxics in large lakes, the failures may not be visible to us in the absence of rigorous validation. Or, they may not be apparent because it will take a long time for these effects to manifest. When looking at the Great Lakes one should ask the question:

> ..."Can the simple toxicity test which we are now using precisely predict toxic effects on the large ecosystems?"...

If the answer is no, then we must then consider what additional methods are needed.

The National Research Council report (1981) calls for concomitant use of the single species tests with community and ecosystem level tests. It explicitly recognizes the value of single species tests as the author has repeatedly done elsewhere. The question is not whether single species tests should be replaced, but rather are they alone sufficient for the kinds of judgments we must make on the effects of toxics in large lake ecosystems.

If we think they are sufficient, at the very least, we should have more scientifically justifiable validation of this assumption, which should be published in the peer-reviewed literature so that the usual processes of scientific judgment about an assumption or hypothesis can occur. If we think they are not adequate, we must begin to develop test methods that will permit us to make predictions of the effects of toxics on large lakes and other ecosystems with greater precision and reliability. Books on this subject are already available (e.g. Hammonds 1981, Cairns 1985) and a number of articles in recent issues of professional journals. There is not likely to be the kind of further development necessary until the need for improved ability to predict the effects of toxics on large ecosystems is clearly identified and acknowledged.

Some industries will require legislation before they will take any move beyond single species testing. Other industries are already gathering much information (e.g. Woltering 1985). The book **Multispecies Toxicity Testing** (Cairns 1985), provides evidence that a number of such tests are presently available, their replicability is much higher than one might have thought intuitively reasonable. A quality assurance program is being developed, and the process for quality assurance is being developed. The requirements of industry (Loewengart and Maki 1985) have been discussed and the regulatory position stated (Tebo 1985, Mount 1985). The fact that a multispecies toxicity testing symposium was co-sponsored by the Ecological Society of America and the Society for Environmental Toxicology and Chemistry shows that both societies feel the problem deserves attention.

There is a curious reversal in burden of proof with regard to the assumption of efficacy and effectiveness for present single species test methods. Critics of the overuse of these methods, for extrapolations beyond their capabilities, are required to prove that they are inadequate, as if all present uses of the single species toxicity tests are adequate until proven otherwise. Innocent until proven guilty is the cornerstone of laws governing human freedom in the United States. However, it is not a cornerstone of scientific method. In science, nothing should be accepted without proof. Instead of regulatory agencies and industry refusing to go beyond single species testing in any significant way until the need for this has been **"scientifically demonstrated"**, science should require that methodologies presently in heavy use be curtailed until their efficacy as used has been demonstrated with scientifically justifiable, not circumstantial, evidence. Unfortunately, quality assurance has focused on the replicability of single species methods in round robin testing (i.e. can a number of laboratories get the same result with the same unknown test material) rather than quality assurance in terms of the way the data are being used (i.e. do we have direct confirming evidence based on hard scientific data that large lakes are indeed protected from toxic effects at concentrations deemed acceptable by single species toxicity tests?).

This reversal of burden of proof is most striking with the acid rain situation. Scientists are being asked to demonstrate, with conclusive scientific evidence, that the clean-up measures being proposed will have the beneficial environmental results claimed for them. In reality, the burden of the proof is on industry to show convincingly and with strong scientifically justifiable evidence that there is no harm from discharges into the environment. We are acting as if the free and unrestrained use of the environment is as much a right as innocence until proven guilty.

There is not enough money in science to keep up with all the types of discharges that could go into the environment. The only sensible way to restrict these for large lakes and other vast ecosystems is to require the dischargers to demonstrate convincingly that there will be no harm. Until we resolve this burden of proof problem and get it where it should be, pollution control will be reactive instead of managerial and action will only be taken after damage is demonstrated.

The author can think of no conference better suited to a crystallization of this problem than the World Conference on Great Lakes. The disparity between the size of the test system and the size of the natural system is striking, the replication of the important cause/effect pathways in the test system is miniscule compared to those existing in the natural system, and finally, the large lakes once damaged will be exceedingly difficult, if not impossible to restore.

It is our responsibility to ensure that the potential effects of toxics are predicted as precisely as science now permits and to develop methods making greater precision available as expeditiously as possible. While this is being done, we should explicitly state the assumptions associated with the use of the single species toxicity test since it is upon them that we now place our primary reliance. The scientifically sound validation process may show that the single species tests provide all the evidence we need. If this situation develops, no one will be more pleased than I! It would solve a complex problem with greater easy than this author would have thought possible. If, on the other hand, there are major deficiencies and weaknesses, the sooner they are identified the better. Above all, the protective measures for large lakes should be based on good science and the author is convinced that this is not yet the case.

This situation should accentuate our awareness of the deficiencies in present estimates of hazard to large aquatic ecosystems:

1. Most test methods are designed for rivers and small lakes.

2. Our methods have been dominated by single species toxicity tests which may be unsuitable for precise estimates of effects at higher levels of biological organization.

3. Toxicity tests may present low environmental realism.

4. Validation of predictions of hazard are exceedingly rare (i.e. poor error control is problematic in predictive methods).

5. Extrapolation (through the use of application factors) to critical ecosystem

thresholds is more common than direct measurement of response.

6.Ecosystem quality control conditions are rarely as explicitly stated as are the physical (e.g. temperature) or chemical (e.g. dissolved oxygen concentration) parameters.

7.Monitoring to ensure that the quality control system is working may be inadequate.

8.The ubiquitous problem of cumulative effects (ranging from repeated removals of organisms and minerals to additions of potentially toxic materials) are not usually addressed in a substantive way.

9.Uncertainty associated with an estimate of hazard is often not explicitly stated in ecological terms.

To improve estimates of hazard due to toxics, the following suggestions are offered:

1.Data for hazard assessment should be gathered in a systematic, orderly way (i.e. protocols).

2.There must be recognition that some toxics require more testing than others.

3.The entire test series should consist of:

i) range finding tests,
ii) predictive tests,
iii) validation of predictions,and
iv) quality control monitoring.

4.Avoid prescriptive legislation that ignores ecosystem differences (e.g. regulations prohibiting acidic discharges from artificial wetlands).

5.Couple environmental fate information with ecological effects information.

6.Involve ecotoxicologists from the beginning (i.e. research and development of the potentially toxic material) since they can advise on potential outcomes and sometimes determine when seemingly incompatible values might be in conflict.

7. Treat projects, where the outcome is highly uncertain, as experiments. This will increase the information base and is more likely to produce useful information than would blind adherence to prescriptive legislation.

8. Publish information in peer-reviewed professional journals. This will make information more readily available and increase confidence in it. Quality control in the **"grey literature"** often leaves much to be desired. In addition there is less opportunity for professional rebuttal. Many journals encourage critiques of papers after they appear and publish these together with the author's response.

9. Uncertainty is unavoidable but pretending it does not exist results in poor hazard evaluation. If uncertainty is dealt with in a straightforward way, managers are better prepared for alternative outcomes.

10. Always validate laboratory test result in natural systems or surrogates that have a higher degree of environmental realism (e.g. heterogenity in space and time) than the laboratory systems.

11. Even when a chemical appears **"safe"** be alert for cumulative effects either with other chemicals or other environmental stresses (e.g. temperature extremes, low dissolved oxygen concentration).

ACKNOWLEDGEMENTS

I am greatly indebted to Darla Donald for preparing this manuscript for publication and to Angela Miller for typing the final draft.

REFERENCES

Cairns, J., Jr. 1980. Guest editorial: Beyond single species testing. Mar. Environ. Res. 4(3):157-159.

Cairns, J., Jr. 1980-81. Guest editorial: Sequential vs. simultaneous toxicity testing protocols. Mar. Environ. Res. 4;165-166.

Cairns,J., Jr. 1981. Biological monitoring. Part VI: Future needs. Water Res. 15:941-952.

Cairns, J., Jr. 1983a. Are single species toxicity tests alone adequate for estimating hazard? Hydrobiologia 100:47-57.

Cairns,J., Jr. 1983b. The case for simultaneous toxicity testing at different levels of biological organization. In: Aquatic Toxicology and Hazard Assessment: Sixth symposium, STP 802. ed. W.E. Bishop, R.D. Cardwell and B. Heidolph, pp. 111-127. Philadelphia, Pa.: American Society for Testing and Materials.

Cairns, J., Jr. ed. 1985. Multispecies Toxicity Testing. New York. Pergamon Press.

Cairns,J., Jr. 1986. Multispecies toxicity testing: a new information base for hazard evaluation. Curr. Prac. Environ. Sci. Eng. 2:185-195.

Cairns, J., Jr. in press (a). The myth of the most sensitive species. Bioscience.

Cairns, J., Jr. in press (b). Keynote address: Politics, Economics, Science - Going Beyond Disciplinary Boundaries to Protect Aquatic Ecosystems. In: Persistent toxic substances and the health of aquatic communities. ed. J. Gannon and M. Evans. New York: Wiley, Inc.

Cairns, J., Jr. in press (c). What is meant by validation? Hydrobiologia.

Cairns, J., Jr. in press (c). What constitutes field validation of predictions based on laboratory evidence? In: Biological assessment of hazardous wastes. 10th Symposium. Philadelphia, Pa.: American Society for Testing and Materials.

Cairns, J., Jr. and Cherry, D.S. 1983. A site-specific field and laboratory evaluation of fish and Asiatic clam population responses to coal fired power plant discharges. Water Sci. Tech. 15:10-37.

Cairns, J., Jr., Dickson, K.L. and Maki, A., ed. 1978. Estimating the hazard of chemical substances to aquatic life. STP 657. Philadelphia, Pa.: American Society for Testing and Materials. 278 pp.

Dicks, B. 1973. Some effects of Kuwait crude oil on the limpet, Patella vulgata. Environ. Pollut. 5:219-229.

Dickson, K.L., Cairns, J., Jr. and Buikema, A.L., Jr. 1984. Comparison of biomonitoring and techniques for evaluating effects of jet fuel on bluegill sunfish (Lepomis macrochirus). In: Freshwater biological monitoring. ed. D. Pascoe and R.W.Edwards, pp. 103-122. Oxford: Pergamon Press.

Hammonds, A.S., ed. 1981. Methods for Ecological Toxicology: A Critical Review of Multispecies Tests. Springfield, Va.: National Technical Information Service.

Herbes, S.E. 1986. Predictive models and field studies of the fate of complex mixtures. In: Environmental hazard assessment of effluents. H.L. Bergman, R.A. Kimerle and A.W. Maki, (eds.) pp. 1720190. New York: Pergamon Press.

Janzen, D.H. 1966. Coevolution of mutualism between ants and acacias in Central America. Evolution 20;249-275.

Kimball, K.D. and Levin, S.A. 1985. Limitations of laboratory bioassays: the need for ecosystem-level testing. BioScience 35(3):165-171.

Krebs, C.T. and Burns, K.A. 1977. Long-term effects of an oil spill on populations of the salt-marsh crab Uca pugnax. Science 197:484-487.

Livingstone, R.J., Diaz, R.J. and White, D.C. 1985. Field validation of laboratory-derived multispecies aquatic test systems. U.S. Environmental Protection Agency, Project Summary 600/S-4-85/039, U.S. Government Printing Office, Washington, D.C. 7 pp.

Loewengart, G. and Maki, A.W. 1985. Multispecies toxicity tests in the safety assessment of chemicals: necessity or curiosity? In: Multispecies toxicity testing. J.Cairns, Jr. (ed.) pp. 1-12. New York: Pergamon Press.

McIntosh, R.P. 1980. The relationship between succession and the recovery process. In: The recovery process in damaged ecosystems. J. Cairns, Jr., pp. 11-62. Ann Arbor Science Publishers, Inc.

Mount, D.I. 1985. Specific problems in using multispecies toxicity tests for regulatory purposes. In: Multispecies toxicity testing. J. Cairns, Jr.,(ed) pp. 12-18. New York: Pergamon Press.

Mount, D.I. and Norberg, T.J. 1984. A seven-day life-cycle cladoceran toxicity test. Environ. Toxicol. Chem. 3;425-434.

National Research Council 1981. Report by the Committee to Review Methods for Ecotoxicology, Testing for effects of chemicals in ecosystems. Washington, D.C. National Academy Press. 103 pp.

National Research Council 1986. Ecological knowledge and environmental problem solving, Report of the Committee on the Applications of Ecological Theory to Environmental Problems. Washington, D.C.: National Academy Press. 388 pp.

Niederlehner, B.R., Pratt, J.R., Buikema, A.L., Jr. and Cairns, J., Jr. 1985. Laboratory test evaluating effects of cadmium on freshwater protozoan communities. Environ. Toxicol. Chem. 4:155-166.

Norberg, T.J. and Mount, D.I. 1985. A new fathead minnow (*Pimephales promelas*) subchronic toxicity test. Environ. Toxicol. Chem. 4:711-718.

Odum, E.P., Fin, J.T. and Franz, E.H. 1979. Perturbation theory and the subsidy-stress gradient. BioScience 29:349-352.

Paine, R.T. 1974. Intertidal community structure. Experimental studies on the relationship between a dominant competitor and its principal predator. Oecologia 15: 93-120.

Petrocelli, S.R. 1985. Chronic toxicity tests. In: Fundamentals of aquatic toxicology. ed. G.M.Rand and S.R. Petrocelli, pp. 96-109. Washington, D.C.: Hemisphere Publishing Corporation.

Svedmark, M., Braaten, B., Emanuelsson, E. and Granmo, A. 1971. Biological effects of surface active agents on marine animals. Mar. Biol. 9:183-201.

Tebo, L.B., Jr. 1985. Technical considerations related to the regulatory use of multispecies toxicity tests. In: Multispecies toxicity testing. J. Cairns, Jr., (ed.) pp. 19-26. New York: Pergamon Press.

Weis, J.S. 1985. Letters to the editor: Species in ecosystems. BioScience 35(6):330.

Woltering, D.M. 1985. Population responses to chemical exposure in aquatic multispecies systems. In: Multispecies toxicity testing. J. Cairns, Jr., (ed.) pp. 61-75. New York: Pergamon Press.

CHAPTER 18

MICHIGAN'S PROCESS FOR REGULATING TOXIC SUBSTANCES IN SURFACE WATER PERMITS

James E. Grant

Michigan Department of Natural Resources, Surface Water Quality Division, P.O. Box 30028, Lansing, Michigan 48909

ABSTRACT

A necessary aspect of fisheries management is a water pollution control program that will not only provide nontoxic water quality conditions but also ensure that fish do not contain unacceptable levels of toxic substances for human consumption. Michigan has recently promulgated revisions to Rule 323.1057 of its Water Quality Standards that establish a regulatory process that will protect public health and the environment from discharges of toxic substances from point source surface water discharges. Rule 57(2) specifically addresses the development of allowable toxicant levels in the waters of the state applicable to point source dischargers. The universe of chemicals to which the subrule applies is defined, an upper boundary on estimated excess risk of 1 in 100,000 for non-threshold carcinogens is established, comprehensive procedural guidelines are mandated and a mechanism for issuance of scheduled abatement permits is provided. This paper reviews the development of the rule amendments and discusses key aspects of the adopted rules and guidelines.

INTRODUCTION

Michigan's unique geographic position at the heart of the Great Lakes has provided it with an enormous fisheries resource potential to manage. Because fish are an integration of their aquatic environment and reflect its quality, certain toxic substances have bioaccumulated in some Great Lakes fish species to unacceptable levels. This has had serious implications in Great Lakes Fish management decisions concerning expenditure of funds for fish development or restoration projects. Therefore, Michigan has recently expanded its surface water pollution control program to provide not only nontoxic water quality conditions, but also to assure that fish are fit for human consumption.

A goal of the surface water permit program of the Michigan Department of Natural Resources (MDNR) is to protect public health and the environment from toxic substances discharged from point sources. Basic elements of the program are the development of technology-based effluent limitations determined from federal Best Available Treatment (BAT) requirements and the development of water quality-based effluent limitations to assure that water quality standards are met. Development of the levels necessary for calculation of the water quality-based limitations is the subject of this paper. It is realized that additional programs to control impacts of toxic substances to the Great Lakes from non-point and aerial sources will also be necessary for protection of the Great Lakes fisheries resource.

LEGISLATION AND WATER QUALITY STANDARDS

The basic water pollution legislation in Michigan is the Michigan Water Resources Commission Act, Act 245, P.A. of 1929, as amended. Section 6(a) reads:

> ..."It shall be unlawful for any persons directly or indirectly to discharge into the waters of the state any substance which is or may become injurious to the public health, safety or welfare; or which is or may become injurious to domestic, commercial, industrial, agricultural, recreational, or other uses which are being or may be made of such waters; or which is or may become injurious to the value or utility of riparian lands; or which is or may become injurious to

livestock, wild animals, birds, fish, aquatic life, or plant..."

Sections 2, 5 and 7 empower the Water Resources Commission (WRC) to control the pollution of the waters of the state through issuance of permits which restrict the constituents of discharges to levels which assure compliance with state water quality standards.

Water quality standards are provisions of law which define the water quality goals of a water body by designating uses and establishing criteria to protect those uses. As used in the surface water discharge permit program, water quality standards serve as the regulatory basis for the establishment of water quality-based controls beyond the technology-based levels of treatment required by the Clean Water Act.

Part 4 of the General Rules of the Michigan Water Resources Commission contains the State Water Quality Standards. Michigan's first formal water quality standards were promulgated in 1967 and revised in 1973. In January, 1985, significant amendments to Rule 323.1057 were adopted. Rule 57 is Michigan's's Toxic Substance Water Quality Standard.

Rule 57 was revised because the 1973 version had been promulgated at a time when the body of knowledge concerning toxic substances was much less than it is today. Outdated literature was cited and its primary emphasis was on aquatic acute and chronic toxicity. The need to have a rule that placed more emphasis on public health was apparent.

The rule 57 revision process was long and controversial and dated back to 1976. The incorporation of a risk assessment process for carcinogens into the rule was a more recent cause of delay. The establishment of a Rule 57 Advisory Committee in 1981 representing outside interest groups was a key element in finally drafting a rule package that could be agreed to by the regulated community and the major environmental groups in Michigan.

RULE 57

Rule 57 is considered a narrative water quality standard as opposed to a numerical rule which would have absolute values specified for a list of toxic substances. Because of the rapid advances in the field of toxicology, difficulties we have had in the past when attempting to promulgate a numerical rule, and the complexities of amending rules in this state, we feel a narrative rule blended with more specific guidelines (as defined by the Michigan Administrative Procedures Act) is the most logical approach.

The rule is divided into two subrules. The first being a general statement prohibiting injurious levels of toxic substances in the waters of the state and indicating that the Commission determines allowable levels by using appropriate scientific data. Determination of allowable levels for situations other than point source discharges needs to be done on a case-by-case basis.

Subrule (2) specifically addresses the development of allowable toxicant levels in the waters of the state applicable to point source discharges. More detail is provided by defining the universe of chemicals to which the subrule applies, placing an estimated upper boundary on risk of 1 in 100,000 for carcinogens not determined to cause cancer by a threshold mechanism, indicating that the allowable toxicant levels apply at the edge of mixing zones, mandating the Rule 57(2) guidelines and establishing a mechanism for issuing scheduled abatement permits.

The Michigan Critical Materials List and EPA'S lists of priority pollutants and hazardous materials are used as the generic lists of concern. However, if a chemical not on the lists is of concern for a specific permit, the Commission can make a determination to include it on a case-by-case basis.

The risk assessment process and upper limit on risk for chemicals assumed to be non-threshold carcinogens was agreed upon by the Rule 57 Advisory Committee and MDNR staff. Rule 57 requires that a point source discharge not create an estimated level of risk to public health greater than 1 in 100,000 above background in the surface water after mixing with the allowable receiving stream volume specified in R 323.1082 (mixing zone rule) and calculated using the model and assumptions specified in the Rule 57(2) guidelines. MDNR staff feel that the actual risk to the public health associated with exposure to these chemicals in most surface waters of the state under these conditions, is considerably less than 1 in 100,000 and is well below that of common everyday risks.

The concept of blending Rule 57 with the Rule 57(2) guidelines provides a more flexible package than if all the details of the guidelines were in rule form. We believe that the procedures are practical and that they are being implicated at this time. However, it is important to realize that the knowledge and understanding of toxic substances is rapidly expanding. The procedures, while valid today, will require periodic review and revision to keep up with the state-of-the-art of the science involved. With this in mind and the fact that guidelines can be changed easier than rules, Rule 57 was kept in the more general narrative form and most of the highly technical detailed procedures were placed in the guidelines. The Rule 57(2) guidelines will be discussed in more detail later in this paper.

Rule 57(2) states that the Commission may issue a scheduled abatement permit if immediate attainment of the allowable level of a toxic substance is not economically or technically feasible and no prudent alternative exists. In addition, the permitted discharge during the period of noncompliance cannot be of a quality which causes long-term adverse impacts to the public health, safety and welfare. These permits are meant to be of an interim nature and must include a schedule to achieve reasonable progress toward compliance with the final limits. During the developmental stages of Rule 57, considerable comments were submitted concerning the possible adverse economic impact of promulgating Rule 57. The facility specific scheduled abatement permit is the most appropriate mechanism to address the economic aspects of the rule.

RULE 57(2) GUIDELINES

The Rule 57(2) guidelines are specifically mandated in Rule 57(2)(d). These guidelines were adopted pursuant to the Administrative Procedures Act and are only binding on the MDNR. They set forth procedures that MDNR staff use in the development of recommendations to the Water Resources Commission on allowable levels of toxic substances in the waters of the state applicable to point source discharge permits. They also set forth the minimum toxicity data needed for a chemical to enable staff to derive their recommendations. Minimum data consists of a rat oral LD50, a 48 hour EC50 for a daphnid, and a 96 hour LC50 for a fathead minnow or rainbow trout.

The guidelines contain detailed procedures for calculating levels necessary to protect aquatic life (Aquatic Chronic Value), wildlife (Terrestrial Life Cycle Safe Concentration), and public health from threshold effect toxic substances (Human Life Cycle

Safe Concentration); and concentrations providing an acceptable degree of protection to public health for cancer, hereditary mutagenic effects or genotoxic teratogenic effects. The most restrictive of the above values is used as the Rule 57(2) level in the surface water after a discharge is mixed with the allowable receiving stream volume. Discussion on the calculation of these values follows:

Aquatic Chronic Value

The aquatic chronic value (ACV) is the highest concentration of a chemical or combination of chemicals which theoretically will produce no adverse effects on important aquatic organisms (and their progeny) exposed continuously for a lifetime. The ACV can be calculated on a chemical specific basis or by using biological techniques, such as bioassays, to assure that chronically toxic conditions do not exist for important aquatic life in the waters of the state. With the chemical specific approach, a specific numerical value is derived for each chemical using the procedures in the guidelines. The procedures also account, as appropriate, for the effects of various water quality characteristics (i.e. hardness, pH) on the toxicity of a chemical substance. Site specific data are preferred and used whenever possible.

The chemical specific mechanism used to calculate the ACV for a toxic substance depends upon the number of chronic data points available for that substance. When six or more appropriate chronic data points are available for a chemical, the ACV is calculated directly from fish and macroinvertebrate chronic toxicity data for that chemical using procedures similar to those described in U.S. EPA, 1985. ACVs for chemical substances calculated using this procedure theoretically are designed to be equivalent to, or less than, the chemical's chronic value for 95 percent of all fish and aquatic macroinvertebrate species resident to Michigan's waters.

Unfortunately, there exists a large number of chemical substances for which there is little or no chronic data available. For these chemical substances, the ACV must be predicted from Final Acute Values (FAV) using appropriate application factors. A FAV corresponds to the highest concentration of a chemical in water which theoretically will kill or significantly impair 50 percent of a population of important aquatic organisms exposed continuously for a short period of time (96 hours for fish and aquatic macroinvertebrates, except 48 hours for cladocerans and chironomids). The FAV is calculated using the modified U.S. EPA 1985 approach when six or more appropriate acute data points are available. If this data base is not available, the FAV is predicted by dividing the most sensitive species tested (rainbow trout/daphnid; or fathead minnow/daphnid) by a species sensitivity factor of 5 if rainbow trout is present in the data base or 10 if absent. These species sensitivity factors were derived by MDNR staff (rationale is available from the author upon request). The ACV is predicted by dividing the FAV by a chemical-specific application factor (acute LC50/chronic value ratios) for those chemical substances which have at least one acute/chronic ratio available. When chemical-specific application factors can not be determined due to an absence of appropriate chronic data, the ACV is predicted by dividing the FAV by a general application factor of 45. This application factor was derived by MDNR staff (rationale available upon request) and corresponds to about the 80th percentile rank of all similarly selected ratios.

The details of using biological techniques or the toxicity-based approach have to be established on a case-by-case basis. The advantages of using this approach are that the interaction of chemicals is inherently addressed by the test, incomplete chemical characterization of the effluent can be accounted for, chemical specific toxicity testing can be reduced in certain cases and a more site specific determination can be made.

Human Life Cycle Safe Concentration

The human life cycle safe concentration (HLSC) is the highest concentration of a chemical which causes no significant adverse effects to humans and their offspring when exposed continuously for a lifetime. HLSCs are derived to provide an adequate margin of safety against the adverse effects of chemicals which elicit threshold responses, excluding carcinogenic effects. The threshold response assumes that an organism has a physiological reserve which must be depleted before an effect is manifested.

To derive an HLSC for a chemical the NOAEL for laboratory animals or man is determined. Although use of human data is preferred in most cases these data are lacking and animal data must be used instead. The NOAEL is then divided by the appropriate uncertainty factor (10-1,000) to determine the acceptable dose for a human. This factor is used to account for the uncertainties in trying to predict an acceptable exposure level for the general human population based upon experimental animal data or limited human data.

For many chemicals, appropriate toxicological data NOAELs are not available to derive an HLSC by this method. In the absence of an adequate toxicity data base, procedures have been developed to derive an HLSC from a single acute toxicity data point, i.e. an oral rat LD50. The procedure for deriving an HLSC from an oral rat LD50 involves the use of an acute to chronic application factor. The acute to chronic applicator factor is a numerical value by which the acute oral rat LD50 is adjusted. The value of this factor as derived by MDNR staff is 0.0001 (rationale available upon request). The oral rat LD50 is multiplied by the acute to chronic application factor (0.0001) and the value obtained from this procedure is used as a surrogate NOAEL.

The acceptable dose or milligrams of toxicant (MgT) is translated into a water concentration using the following formula:

$$HLSC = \frac{MgT\ (mg/day)}{WC + (F * BCF)}$$

where: MgT = milligrams of toxicant/day
WC = volume water consumed/day
F = fish consumed per day
BCF = bioconcentration factor of chemical

For all surface waters, it is assumed that a person consumed 6.5 grams of contaminated fish per day (approximately 5 lb/yr) for a lifetime. This value is based upon a USEPA survey of fish and shellfish consumption in the U.S..

The volume of water consumed per day is assumed to be an untreated 2.0 litres for surface waters protected as a drinking water source, and an untreated 0.01 litres for all other surface waters. The value of 2.0 litre was recommended by the USEPA for establishing drinking water standards. The value of 0.01 litres of water per day for surface waters not protected for drinking water is to account for incidental exposure such as absorption through the skin or ingestion of small quantities of water while swimming or using the waters for other recreational purposes.

Terrestrial Life Cycle Safe Concentration

The purpose of establishing terrestrial life cycle safe (TLSC) values it to determine surface water concentrations which are considered acceptable for the wildlife and livestock that utilize these waters. The TLSC is defined as the highest aqueous concentration of a toxicant which causes no significant reduction in the growth, reproduction, viability or usefulness (in the commercial and/or recreational sense) of a population of exposed organisms (utilizing the receiving waters as a drinking water source), over several generations.

To derive a TLSC, the scientific literature regarding the toxicological effects of a chemical is reviewed to determine a no observable adverse effect level (NOAEL) for appropriate mammalian and/or avian organisms. Data on organisms native to Michigan and likely to be utilizing the particular surface water are preferred for calculating the TLSC. In most cases, however, such data are lacking and the data from common laboratory animals (usually rodents) must be used instead. The experimental NOAEL is then divided by an uncertainty factor ranging from 10-100. This uncertainty factor is to account for 1) species variability, since data from one species are used to predict an acceptable level for all wildlife, and 2) inadequacies in study designs or availability of data. When appropriate NOAEL data are not available, a TLSC may be calculated from an oral rat LD50 by the following equation:

$$TLSC = \frac{LD50(mg/kg) * \frac{Wa}{VW} * M}{10}$$

where: Wa = weight of test animal (kg)
VW = volume of water consumed by test animal per day
M = acute to chronic application factor of 0.0001 derived by MDNR staff (rationale available upon request).

Cancer Risk Value

Due to the limits of current predictive testing, the Rule 57 guidelines make the conservative assumption that any chemical which has been shown to be carcinogenic in one animal bioassay of good quality, is a complete carcinogen having no threshold. However, the guidelines do include a mechanisms for evaluating a carcinogen on a case-by-case basis if the preponderance of data suggests the cancer is caused by a threshold mechanism and does not interact with DNA. A committee of scientists expert in the field of carcinogenesis may be convened when MDNR staff will benefit from advice and recommendations on this issue, or other highly technical scientific issues, which require additional technical expertise to resolve.

If appropriate human epidemiological data are available, an extrapolation from high doses is necessary in order to estimate the carcinogenic risk for the chemical at low concentrations. There are no standard guidelines available to estimate the risk from human epidemiology studies. However, the use of adequate human exposure data to estimate the risks associated with a carcinogenic chemical is a preferred method and when necessary, the MDNR may convene an expert committee to advise staff on an appropriate methodology in order to utilize these data.

When human epidemiological evidence is not available, the carcinogenic risk to humans is extrapolated from experimental animal data. There is no conclusive scientific evidence for the choice of one mathematical model over another; however, the linearized multistage model, is used since no other extrapolation model has as much regulatory acceptance. Use of the upper 95 percent confidence limit to estimate the dose rather than extrapolation from the maximum likelihood estimate dose gives a more stable value which does not change appreciably with minor variability in the biological response at the lower doses. The use of this methodology provides a plausible upper limit of risk.

The Rule 57 Advisory Committee felt that the risk associated with exposure to these chemicals in ambient water should generally be below that of common everyday risks and recommended that an estimated risk level of 1 in 100,000 be used as the upper boundary of risk for establishing allowable levels of carcinogens in the waters of the state applicable to point source discharges. MDNR staff agree with this recommendation. Greater levels of protection are also evaluated at facilities where achievable through utilization of control measures already in place.

Cancer risk values are calculated using the following formula:

$$C = \frac{D * 70\text{ kg}}{WC + (F * BCF)}$$

Where: D = dose (mg/kg.day)
WC = water consumed/day
F = fish consumed/day
BCF = bioconcentration factor for the chemical

The values are calculated on the basis of a 70 kg human and the fish and water exposure assumptions are the same as those used for the HLSC values.

Hereditary Mutagen and Genotoxic Teratogen Values

The levels providing an acceptable degree of protection to public health for hereditary mutagenic effects and genotoxic teratogenic effects are derived by MDNR staff on a case-by-case with help, as needed from an expert committee of scientists.

Water quality standards are to protect the public health and welfare, enhance the quality of water and serve the purposes of the Clean Water Act. MDNR staff feel that public health and the environment are protected with an adequate margin of safety by developing water quality-based discharge levels for toxic substances in the surface water discharge permit program based on Rule 57 and the Rule 57(2) Guidelines. Copies of the Michigan Water Quality Standards and the Rule 57(2) Guidelines are available from the author upon request.

ACKNOWLEDGEMENTS

The author acknowledges the support from Michigan's regulated community and major environmental groups and the efforts of the Toxic Chemical Evaluation Section staff and the Rule 57 Advisory Committee members in the development of this process.

REFERENCES

EPA (Environmental Protection Agency) 1985. Guidelines for deriving numerical national water quality criteria for the protection of aquatic organisms and their uses. U.S. Environmental Protection Agency, NTIS Number PB 85-227049, Environmental Research Lab, Duluth, MI.

Crump, K.S. and W.W. Watson 1979. GLOBAL 79. A FORTRAN program to extrapolate dichotomous animal carcinogenicity data to low doses. National Institute of Environmental Health Sciences Contract NOI-ES-2123.

Index